NEW DIRECTIONS
FOR METHODOLOGY OF
SOCIAL AND
BEHAVIORAL SCIENCE

Number 7 • 1981

NEW DIRECTIONS FOR METHODOLOGY OF SOCIAL AND BEHAVIORAL SCIENCE

A Quarterly Sourcebook
Donald W. Fiske, Editor-in-Chief

Number 7, 1981

Biopolitics:
Ethological and
Physiological Approaches

Meredith W. Watts
Guest Editor

Jossey-Bass Inc., Publishers
San Francisco • Washington • London

BIOPOLITICS: ETHOLOGICAL AND PHYSIOLOGICAL APPROACHES
New Directions for Methodology of Social and Behavioral Science
Number 7, 1981
Meredith W. Watts, Guest Editor

New Directions for Methodology of Social and Behavioral Science
is published quarterly by Jossey-Bass Inc., Publishers.
Subscriptions are available at the regular rate for institutions,
libraries, and agencies of $30 for one year. Individuals may
subscribe at the special professional rate of $18 for one year.

Correspondence:
Subscriptions, single-issue orders, change of address notices,
undelivered copies, and other correspondence should be sent to
New Directions Subscriptions, Jossey-Bass Inc., Publishers,
433 California Street, San Francisco, California 94104.

Editorial correspondence should be sent to the Editor-in-Chief,
Donald W. Fiske, University of Chicago, Chicago, Illinois 60637.

Library of Congress Catalogue Card Number LC 80-84294

International Standard Serial Number ISSN 0271-1249

International Standard Book Number ISBN 87589-851-3

Cover design by Willi Baum

Manufactured in the United States of America

Contents

Preface

The field of methodology is not restricted by the boundaries around any single substantive discipline. Its principles, concepts, and techniques have wide applicability, even though some specific methods of investigation may be used primarily within a single field. Each of the sourcebooks in this series is devoted to a topic pertinent to investigation in more than one social or behavioral science.

The chapters in this sourcebook are concerned with "applying theories and methods from the contemporary life sciences to social and political behavior," as the guest editor writes in his introduction. The emphasis on political behavior serves to integrate the several chapters. That emphasis, however, should not lead the reader to overlook the broader implications of these contributions. Two theses in this volume have wide applicability: the use of ethological and other biological conceptualizations and methodologies in social science and, more generally, the borrowing and adapting of concepts and methods from one discipline for work in another discipline studying rather dissimilar phenomena.

Donald W. Fiske
Editor-in-Chief

This chapter provides an overview of biopolitics — the fast-growing tradition of applying theories and methods from the contemporary life sciences to social and political behavior. The editor presents an overview of the area and its particular problems and methodologies and introduces the chapters in this volume.

Editor's Notes and Introduction

Meredith W. Watts

The chapters in this volume share in a tradition that is fast-growing — that of applying theories and methods from the contemporary life sciences to social and political behavior. This introduction presents both an overview of the area and the particular problems and methodologies that it favors before it briefly introduces the chapters that make up this volume.

Biopolitics

The synthetic term *biopolitics* refers to an interstitial discipline with an identifiable set of intellectual concerns, methodological interests, and a growing number of scholars. Where, in the period from 1963 to 1969, there was a total of twenty-one written documents clearly in the biopolitical tradition, there were ninety-one in 1970–1974 and 176 between 1975 and 1979 (Somit, Peterson, Richardson, and Goldfischer, 1980). One of the first papers appeared in 1964 (Caldwell, 1964), and reviews of the literature have appeared every few years since then. Those interested in the area, though hardly identifiable with sociobiology per se, have been noticed and picketed by the critics of sociobiology — a rare tribute to the imagined or predicted impact of biopolitics. Perhaps most important, there has emerged in the biopolitical literature a recognition that it is time for the rigorous development of empirical research (Wahlke, 1976, p. 258). The appearance of the first biopolitics text (Wiegele, 1979) is another sign of growing vigor.

These chapters examine the nature of some relevant empirical methods for the portion of biopolitics that concerns itself with ethological and physiological approaches to questions of sociopolitical relevance. Of the two approaches, the former receives more emphasis because many of the chapters in this volume were presented as papers at a panel of the International Society of Political Psychology chaired by the editor. Strictly theoretical and narrowly physiological topics were not heavily represented on that panel in order to emphasize a restricted set of themes—in this case, the relevance of ethological methods to the study of aggressiveness, dominance-submission in human groups, and the stress induced by sociopolitical behavior. The object is not only to elucidate a set of methods but also to provide sufficient substance so that others may be encouraged to investigate further the topics represented.

Like the social sciences, which have always made forays into cognate disciplines for ideas and methods, work in biopolitics has often been eclectic. Yet, while this work has been diverse, much of substance, conceptual meaning, and methodological utility has been captured. A brief overview of the emerging conceptualization of the field will provide some perspective. But first, some criticisms of conventional social science, and especially of political science, that motivate the current movement should be described.

As Albert Somit remarked in his review of the literature nearly a decade ago, "With few exceptions, social scientists trained after the First World War simply took it for granted that they could safely ignore man's genetic legacy" (Somit, 1976, p. 293). Glendon Schubert proposed changing this situation by extending the perceived bounds of interdisciplinary relevance for political scientists to include the life sciences: "Behavioral science has not hesitated to take advantage of what additionally could be learned about human behavior by adopting technologies that enhance the capacity to observe organisms in action. What remains to be learned will be primarily the previously unobservable, and this will consist mostly of events that take place within the human body or of human organismic behavior in the context of environments that previously were inaccessible to observation (like the long political past that antedates the invention of writing . . . " (Schubert, 1976, p. 165). The major area of development in theory and method should take place at the intersection of the social sciences and such life sciences as microbiology, ecology, neurophysiology, comparative psychology, ethology (particularly primatology), and genetics. As others already had, he rejected the largely rationalistic, culturally deterministic outlook of the contemporary social sciences, to which many scholars now associated with the biopolitics movement contributed in the past, arguing that (p. 156) "the chief task confronting the political science of the present and future will be to study in-depth interactions between the rational and cognitively determined political actions hypothesized by political behavioralism and the *biology of human organisms which supports and constrains the possibility of any and all political behavior*" (emphasis added).

For the policy scientist who wishes to apply the findings of the social sciences to the study of public policy, the application of such sociopolitical therapy is not aided by the arbitrary deletion of the biological realm of life from our analytical models. This suggests that "our expectation of the probable pay-off for any therapy based on substantial ignorance of underlying causes (to say nothing of direct and indirect effects) ought to be something close to zero improvement" (Schubert, 1976, p. 166).

There is no unanimity on the exact goals of a biopolitics, but there is agreement that a variety of biological variables, whether from ethology, physiology, or elsewhere, must be an operational part of the research designs of social scientists. Masters (1976, p. 199) argues further that biology, rather than physics, should provide the epistemological norm for social scientists: "Like biology—and unlike classical physics—the social sciences study *populations* of organisms that change over time. Like biology—and unlike classical physics—time is an essentially irreversible variable of decisive importance in most of the phenomena analyzed by political science. Like biology—and unlike classical physics—some form of teleological or functional reasoning seems inherent in political life. Finally, like biology—and unlike classical physics—political science studies complex systems (human societies) which are self-replicating organizations of information" (1976, p. 199). Masters notes that biologist G. G. Simpson was himself in fact "content to define the social sciences as those branches of biology dealing with organisms that have language."

Obviously, something unusual has occurred when political scientists of both classical and behavioral traditions find themselves in general agreement on the need for additional attention to biological or life sciences. There may be an internal contradiction in this new synthesis (apologies to Marx and Hegel). The more empirically oriented may prefer experimental and highly quantitative approaches, while the more classically minded may lean toward the teleological implications ("functional logic") of ethology and the survivalist "inclusive fitness" criterion of contemporary evolutionary theory. But these differences in emphasis have tended to be mutually supportive in recent years.

But if such a synthesis is to seek a hearing from the academic community, it must, at the least, draw upon known and familiar concepts and analogues in the disciplines as commonly perceived and accepted (that is, conventional practitioners should recognize the subject matter); bring to bear theoretical and methodological tools that are either new or novel in application; make some reasonable claim, based on evidence, that its insights and empirical findings are different from those forthcoming from the prevailing approaches; and provide a reconceptualization of the subject matter in a manner that links the conventional and the novel in a meaningful way (that is, the new must come from and supplement the old if it is to have impact and acceptance). The chapters in this volume speak in various ways to these criteria.

Has the biopolitics approach produced any reconceptualizations of the phenomenon of politics that differ from the norm? This, I suppose, is a rough first test of whether new and useful concepts are being introduced. The following quotations suggest some of the diversity in this area. As one recent writer stated, "From an ethological perspective, politics can be defined as behavior which simultaneously partakes of the attributes of dominance and submission (which the human primate shares with many other mammals) and those of legal or customary regulation of social life (characteristic of human groups endowed with language)" (Masters, 1976a, p. 197). According to this writer (p. 207), from the biopolitical perspective, "political life is the arena of 'agonistic' (aggression and dominance-oriented) behavior directed to the establishment, maintenance or change of social rules."

Peter Corning, taking an evolutionary perspective that emphasizes species survival, argues, "Politics . . . is the process by which human societies go about cybernating their behavior—the process by which man attempts to cope with 'public' problems, make authoritative decisions for the group, and organize and coordinate the behavior of his fellows. . . . [P]olitics involves the authoritative selection and implementation of society's collective survival strategy" (Corning, 1976, p. 131). Emphasizing the origins of political behavior in prehuman history, including that of protohominids, "the biocybernetic model of politics views political organizations and collective decision-making as an integral part of the human survival enterprise, with behavioral precursors that in fact predate the emergence of man" (Corning, 1976, p. 139; see also Corning, 1973).

Though these statements may not elicit agreement from all the authors of chapters in this volume, they do indicate that biopolitics seeks to add to the range of concepts and definitions with which social and political science should be concerned. Masters (1976a) leans toward ethology and telelogical explanation. Corning posits survival as the goal of political science and draws theoretical explanation from evolutionary biology rather than from ethology or physiology per se. Adopting a more empirical perspective, Schubert rejects the notion that political life began with the polis or, alternatively, with naturalistic fables, as in Hobbes or Rousseau, and argues that "the roots of political behavior go back not thousands but millions of years" (Schubert, 1976, p. 164).

Linking these and other varying conceptions of politics is the moderate and probably unassailable position of Wiegele: "At the present state of the development of the discipline, we can characterize ourselves as having ignored an enormous amount of information about humanity and its real nature that is its nature *as it is lived*. . . . We must adopt a more comprehensive definition of human nature, one that includes *in an operational way* the biological as well as the rational and psychological" (Wiegele, 1979, pp. 144–145, emphasis added). The call for operational inclusion of biological variables in our research designs as opposed to metaphorical or analogical use—is critically important if biopolitics is to produce findings of theoretical and empirical importance.

Varieties of Biopolitics

We can begin by rejecting the false antinomy of humans as nonmaterial, cognitive, rational organisms versus human as animals—to be specific, as "higher" primates. The conceptions are complementary and should be combined in research designs to the fullest possible extent. Although there is no single intersubjectively agreed upon definition of biopolitical research, several components have come to be recognized. Somit originally proposed that four categories could describe usefully the general topography of biopolitics. More recently, he and his associates have proposed the case for more biologically-oriented political science, ethological and sociobiological approaches, biological techniques as research tools, public policy implications, and metaphoric use of biological concepts (Somit, Peterson, Richardson, and Goldfischer, 1980). Items in the first category are discussed above, and the references there lead to a wealth of hortatory and critical literature. Moreover, topics in the first and last categories are combined, as when critical analysis of the social science literature is melded with metaphorical use of evolutionary or ethological theory.

Certainly, political scientists are not alone in their critical and metaphorical essays. Recently, a sociologist combined the neo-Darwinian evolutionary paradigm of systematic adaptation with contemporary social learning theory to produce a synthetic theory of sociocultural evolution (Langton, 1979). Earlier, Donald Campbell dealt with many related issues (1965; 1975) in sociocultural evolution, biology, and psychology. Pribram (1979) crossed several academic boundaries by combining, in a wide-ranging intellectual approach, sources ranging from Merleau-Ponty and B. F. Skinner in his own neuropsychological research. The result, though perhaps appearing unorthodox to specialists in any of the three academic fields from which it has been composed, is yet congenial in spirit and substance to biopolitics. He argues, "Existential-phenomenological psychologies, if they are to attain precision, must enact form in reasonable *structures* that explicate experience. Holonomic, that is, holistic, psychology depends on discovering *transformations* for its precision. By specifying the transfer functions involved in moving from one state to another, the holistic approach is made as scientifically respectable as any other. Explicitly adding structure and transformation to the search for causes is long overdue and imperative if scientific conceptualizations are to deal with the richness of problems raised by the advances in scientific technology" (Pribram, 1979, p. 71).

Rather courageously, Pribram is demanding structure of the phenomenologists and transformational logic of the holists—both of which can be supplied, he argues, by a more complete understanding of the processing system of the central nervous system and, more particularly, of the brain. One need not agree with Pribram to see that his logic is parallel and complementary to that of the biopolitics movement. However, although he cites brain research as the integrator, many in biopolitics would argue that attention to the autonomic and peripheral nervous system is also vital. The latter systems, the object of measurement by psychophysiologists, have been used in social science for some time, but now they are being theoretically integrated into a life sciences approach rather than used simply as indicators.

The third and the fifth items in Somit's classification of biopolitics will not be discussed in detail in this volume. Readers interested in physiological influences on political behavior may consult Schwartz's studies (Schwartz, 1974, 1976; Schwartz, Garrison, and Alouf, 1975) as a starting point and examine the bibliography and literature review produced by Somit and his colleagues (Somit, Peterson, Richardson, and Goldfischer, 1980). Those interested in the public policy implications of biological knowledge will encounter a vast literature both in the social sciences and in more specialized science journals. The Somit, Peterson, Richardson, and Goldfischer bibliography (1980) and Wiegele's (1979) text can provide starting points for the literature produced by political scientists. In all areas mentioned, the reader may wish to consult, in addition to the various bibliographies mentioned, the many inter-disciplinary journals that Wiegele catalogues (Wiegele, 1979, pp. 152–153). Notable among these journals are *Aggressive Behavior,* the *Journal of Social and Biological Structures, Ethology and Sociobiology, Behavioral Ecology and Sociobiology,* and *Behavioral and Brain Sciences.* Less formal but no less useful sources are the *Human Ethology Newsletter* published at the University of Tennessee and *NOTES,* a newsletter published by the Center for Biopolitical Research at Northern Illinois University.

Of Somit's six categories, ethology receives the most substantial treatment in this volume because of the fortuitious opportunity to bring together the critical and theoretical overview of Schubert, the comparative and methodological work of Strayer, Barner-Barry's focus on "authority" among children, and Masters's application of "attention structure" theory to political research. As a group, these papers provide a rich picture of some of the work now being done. They all attempt to transcend metaphors and loose analogies, and their degree of success should give a rough idea of where selected areas of biopolitics now stand.

Ethological and Physiological Approaches to Sociopolitical Phenomena

In turning to the actual studies, a certain division is apparent; namely, between ethological and physiological (or psychophysiological) approaches. This is not exhaustive of the many approaches now being taken in biopolitics, but it does characterize much of the current empirical research. By shunning metaphor, work in each category becomes subject to a variety of academic particularities and afflictions — some new to social science, some not. One old, familiar malady is reductionism; another is analogism. Homologism is less familiar to social scientists, though it is well known in ethology and comparative animal studies. The authors of the chapters in this sourcebook offer their own positions concerning such methodological matters, but a nonempirical example may serve as a useful illustration.

Willhoite (1971; 1975; 1976; 1977) has examined ethology, particularly primatology, and suggests that dominance and deference may be innate characteristics in human society, just as they seem to be in many other primates. So viewed, the political system may strongly reflect one basic attribute of the human species, albeit one that is enormously elaborated by human linguistic and cognitive endowments. For example, hierarchy, xenophobia, and particularism toward one's group appear to have deep roots in the protohuman origins of the species.

Although Willhoite views these statements as subject to empirical testing, they seem to derive from two major lines of argument. The first is that humans are in some sense like the primate species that have been studied by ethologists. This, of course, is argument by analogy, and, like other analogistic theorizing, there is an element of common-sense truth testing applied by the theorizer; more specific empirical testing is necessary to establish the appropriateness of the analogy and to define its parameters. This is by no means a simple exercise; the literature of biopolitics—and, to a much larger extent, the literature of sociobiology—is rife with speculations on whether the human species is more analogous to various species of hunting *candidae,* army ants, large and small rodents, various macaque species, the great apes, or some other species. Although various popularizations of these themes have made only a slight impression on serious scholarly research, awareness of the applicability to humans of research methods originally developed in ethology and other life sciences is increasing.

Arguments based on the great apes often go farther than simple analogy. Such arguments often claim that humans are related through some allopatric ancestor to those primates and that homology (evolutionary relatedness) exists. This is a stronger form of theorizing than analogy but it demands close scrutiny of archeological and anthropological data on the protohuman origins of the species. Empirical data based on homologistic reasoning is not likely to be forthcoming from political scientists; they and other social scientists will probably tend to generate notions that are derivative of the primatologists. But, as discussion of the theoretical felicity of analogy and homology goes on, more social scientists will receive training in ethology and primatology in preparation for research on more than one species. (Strayer's chapter in this volume provides one such example.)

Sometimes defined as "the study of animal behavior in an evolutionary perspective" (Somit, 1968; Wiegele, 1979), classical ethology has generally dealt with naturalistic observations of animals in free or near-free habitats. Although there have been many experimental studies of such phenomena as vacuum behavior (most of which studies are ethically or pragmatically impossible on humans), classical methodology was often characterized by inductive logic, post hoc conceptualization, and reliance on subjectivity of perception in observations of various species (for example, Lorenz, Tinbergen). More recent ethological approaches have focused more on systematic observation, increased instrumentation and audiovisual recording devices, and development of behavior sampling methodologies (Lamb, Stephenson, and Suomi, 1979). Applied to humans, ethology has a few points of tangency with interaction process analysis (Bales, 1970), but its theoretical basis and instrumentation are radically different. For example, the latter approach has a strong dependence on verbal behavior and begins from a presumption that human verbalizations are of primary explanatory power, while ethology typically does not.

Only in a few areas of social research (see Harper and Wiens, 1979; Ekman, 1979; Ekman and others, 1972) has application been made to political phenomena in the nonverbal mode (Frank, 1977; Hermann, 1979). Yet, by stripping away specific verbal content, albeit temporarily, attention can be turned to characteristics of behavior that are less bound to (and studied through) verbalization. Nonverbal behavior patterns may well be the most typical of our species, and many may be shared with other species—particularly those akin to us on the phylogenetic scale.

The chapter by Schubert makes a case for the use of ethological methods in political analysis and gives, along the way, an introduction to literature that is extremely valuable for understanding both the classical and modern ethological traditions. One can see from the relative incidence of the word *behaviour* in Schubert's sources that the impetus is modern ethology is largely non-American. This situation seems to be changing, although American ethologists and primatologists have tended more than their continental counterparts to deal with provisioned and confined rather than feral animal populations.

The relative primacy of theory is also a matter of concern. An emphasis on theory as opposed to empiricism unhampered by prior conceptualization is held in generally high regard by these authors. Schubert makes a clear case for the former, as does Strayer in a sharply focused discussion of dominance and aggression in stable groups of young humans. Though Schubert and Strayer have done considerable research on sentient beings, their interest in the nonverbal aspects of interactions is in a decidedly different analytic mode than the theoretical translations made by Willhoite to macropolitical propositions. With these examples in mind, we can return to the methodological features of analogy and homology, asking what sort of theorizing bridges the animal/human distinction that has been anathema to so many traditional social scientists and to political scientists in particular.

Mario von Cranach has provided a basic set of readings (1976) that helps to elucidate the logical underpinnings of ethological reasoning for the study of humans. In a line of reasoning familiar to social scientists, he suggests that "if two systems have something in common without being related, conclusions are drawn on the basis of analogy" (von Cranach, 1976, p. 2). No strangers to analogical thinking, social scientists have often found human society to be like something else—which might be a computer or advanced information-processing system (for example, Corning's self-cybernating systems, living organisms, or rationalistic/algorithmic problem solvers). In fact, there is a long list of such examples that resembles a Miller Analogies Test for professional certification in the social sciences.

For popular writers such as Ardrey and Morris and for such popular-
ized ethologists as Lorenz and Eibl-Eibesfeldt, there has been the trenchant
criticism that they know too little about human society to make analogical
inferences. The debate over innate versus learned aggressiveness of humans is
a case in point. Those tending toward the former view cite numerous examples of aggressiveness and dominance/submission is nonhuman species as
analogous to human society, while their critics typically deny the analogical
validity of such comparisons. The distinction between the camps is unclear
and shifting, however, because there is conflicting empirical evidence on the
aggressiveness of various primates. (Chimpanzees observed by Goodall in the
Gombe Reserve, for example, have shown cannibalistic and deliberate aggression on occasion.) So, the substantive matter does not resolve itself simply.
However, those who would tread among ethologists should keep in mind the
root significance of the analogical method in ethological inference.

Certainly, Willhoite's theoretical suggestions draw heavily on analogy;
so, too, do the ethologically based chapters in this sourcebook. But they all
have a general orientation toward demonstrating, rather than positing, the
dimensions of analogical validity. Roger Masters, who is represented in this
volume, has elsewhere provided an extended discussion of this logical mode
(Masters, 1976b), and von Cranach describes it as follows: "If the comparison
somehow, as a presupposition or as a goal, comprises the notion of relationship, the assumption that structures of both systems are based on information
that has been passed from one system to the other, or from a common ancestor
of both, similarity between stuctures is considered based on *homology*" (von
Cranach, 1976, p. 2). While contemporary political scientists may not be
infatuated with homology for reasoning from primate social organization and
behavior to those of humans, many social and behavioral scientists are—particularly those anthropologists who watch closely the archeological evidence of
protohuman and prehuman origins. In fact, much of the appeal of ethological
and primatological research is precisely that we can find analogies between
human and other primate behavior and that we are in some manner related to
other primates in evolutionary history.

Barner-Barry's paper discusses the possibilities of longitudinal research
on sociopolitical relationships using an ethological perspective. Her own
empirical research (cited in her reference section) has employed ethological
models (M. R. A. Chance's attention theory, for the most part) in dealing with
children. As others have found (Strayer, Blurton-Jones), the young of the
human species are easier to provision and institutionalize for periods of observation, their behavior is less verbally elaborated and subtle, and their domin-
ance/submission relations are more amenable to analogical comparison with
other primates. Strayer in particular has studied macaques, rhesus monkeys,
and chimpanzees in addition to human children. Roger Masters similarly
derives his approach from contemporary ethological thinking (again, primaily
from the attention theory of Chance, but applying observational and data–cod-

ing techniques derived from human study) to examine the visual imagery conveyed in the media of American presidential candidates. From such work as this, demonstration of the limits of analogical thinking is likely to come. This is a far cry from various uses of intermixed homology and analogy in the popular literature.

The best approach to such inferential matters is, presumably, more and better empirical research. Without presuming to interpret further the authors in this volume, I propose that a close reading of Schubert, Barner-Barry, Strayer, and Masters will give a good overview of the applicability of ethology and provide some research findings that point to the potential of the approach as applied to humans.

Research on the physiological and psychophysiological aspects of sociopolitical behavior has a much longer tradition in the social sciences. By definition, psychophysiology deals with the response of an individual organism to environmental stimuli that are often social. Tursky and his associates (1976) make a strong case for the contemporary use of such techniques in a multimodal approach, although the Lodge and Wahlke study (1972) represents the first contemporary political science study. The tradition is older, however, as evidenced by Harold Lasswell's use of electrodermal monitoring of psychoanalytic interviews during the 1930s. Two and a half decades later, psychophysiology was employed by Rankin and Campbell (1955) and others to study racial attitudes. The area has burgeoned so significantly that there is now a journal called *Psychophysiology,* in which technical and theoretical studies in the field are regularly available.

Research by Tursky, Lodge, and their associates represents the best-developed and technically best-supported work now being produced by political scientists (Foley and others, 1976), although others (Jaros, 1976; Watts, this volume; Watts and Sumi, 1976) have employed psychophysiological methodology. Those sources, as well as the paper by Tursky and others (1976), will lead the reader to a rich literature in social psychophysiology and the arcana of electrodermal, cardiovascular, ocular, and muscular measurement of activity in the autonomic and peripheral nervous systems. The brief chapter by Watts describes some recent use of psychophysiological methods to examine the influence of individual differences in attitude and personality on the responsiveness of persons to audiovisually modeled social behavior.

Also related is the linking of physiological measurement, ethological matters of dominance/deference posturing, and immediate sociopolitical concerns in the paper by Hirsch and Wiegele. Their study uses a technique based on the frequency modulation of human vocalization to determine the relative degree of stress associated with the verbal content. This procedure links the biological substrate of behavior with the explicit verbal content—a linkage not immediately available in many biopolitical approaches. The Psychological Stress Evaluator (PSE) has, like other biopolitical and sociobiological techniques, received critical examination, as the authors point out. The important question, however, is whether the technique is valid for social science research and whether the inferences made therefrom are legitimate. Contention over the validity of voice stress analysis for lie detection is only peripherally relevant to the uses that Hirsch and Wiegele propose and demonstrate, and the relationship between the PSE and criminology should not be allowed to interfere with the potential contributions of the method.

Common to all the chapters in this volume is an interest in the evolutionary history and the biological substrate of human sociopolitical behavior. It is by no means implied that any of these phenomena can be reduced to purely biological concepts. In fact, it is not argued that any of these phenomena is biologically determined; one needs only to accept that there is a biological component and that social behavior has biological parameters. This minimal acceptance is all that is needed for general social scientists to be able to take notice of these developments and consider their possible contributions.

Biopolitics does not attempt to displace any existing approach but rather to provide the social sciences with the theoretical and empirical richness of the life sciences perspective. Its success will very likely be determined, not by the displacement of some current approach, but from integration with contemporary and conventional usage. For example, if we assume that humans are both rational, cognitive creatures and biological entities with appetites and needs, then the integration of biological with cognitive, phenomenological, and behavioral perspectives is more than a homiletics—it is a theoretical and empirical probability.

References

Bales, R. F. *Personality and Interpersonal Behavior.* New York: Holt, Rinehart and Winston, 1970.

Barner-Barry, C. "An Observational Study of Authority in a Preschool Peer Group." *Political Methodology,* 1977, *4,* 415–449.

Caldwell, L. K. "Biopolitics: Science, Ethics, and Public Policy." *The Yale Review,* 1964, *54* (1), 1–16.

Campbell, D. T. "On Conflict Between Biological and Social Evolution and Between Psychological and Moral Tradition." *American Psychologist,* 1975, *30,* 1103–1126.

Campbell, D. T. "Variation and Selective Retention in Sociocultural Evolution." In H. R. Barringer, G. I. Blanksten, and R. Mack (Eds.), *Social Change in Developing Areas: A Reinterpretation of Evolutionary Theory.* Cambridge, Mass.: Schenkman, 1965.

Corning, P. "Politics and the Evolutionary Process." In T. Dobzhansky and others (Eds.), *Evolutionary Biology.* New York: Appleton-Century-Crofts, 1973.

Corning, P. "Toward a Survival-Oriented Policy Science." In A. Somit (Ed.), *Biology and Politics.* The Hague: Mouton, 1976.

Ekman, P. *The Face of Man: Expression of Universal Emotions in a New Guinea Village.* New York: Garland Press, 1979.

Ekman, P., and others. "Facial Expressions of Emotion While Watching Televised Violence as Predictors of Subsequent Aggression." In G. A. Comstock, E. A. Rubinstein, and J. P. Murray (Eds.), *Television and Social Behavior.* Vol. 5: *Television's Effects: Further Explorations.* Washington, D.C.: U.S. Government Printing Office, 1972.

Foley, H., and others. "Evaluation of the Cognitive Component of Political Issues by Use of Classical Conditioning." *Journal of Personality and Social Psychology,* 1976, *34* (5), 865–873.

Frank, R. S. "Nonverbal and Paralinguistic Analysis of Political Behavior: The First McGovern–Humphrey California Debate." In M. G. Hermann and T. W. Milburn (Eds.), *Psychological Examination of Political Leaders.* New York: Free Press, 1977.

Gray, J. A., and Nebylitsyn, V. D. (Eds.). *Biological Bases of Individual Behavior.* New York: Academic Press, 1972.

Harper, R. G., and Wiens, A. N. "Nonverbal Behaviors as Unobtrusive Measures." In L. Sechrest (Ed.), *New Directions for Methodology of Behavioral Science: Unobtrusive Measurement Today,* no. 1. San Francisco: Jossey-Bass, 1979.

Hermann, M. G. "Indicators of Stress in Policymakers During Foreign Policy Crises." *Political Psychology,* 1979, *1* (1), 27–46.

Jaros, D. "Political Conflict and the School Teacher: An Experimental Analysis." *American Journal of Political Science,* 1976, *20,* 327–348.

Lamb, M. E., Stephenson, G. R., and Suomi, S. J. (Eds.). *Social Interaction Analysis: Methodological Issues.* Madison: University of Wisconsin Press, 1979.

Lang, P. J. "A Bio-Informational Theory of Emotional Imagery." *Psychophysiology,* 1979, *16* (6), 495–511.

Langton, J. "Darwinism and the Behavioral Theory of Sociocultural Evolution: An Analysis." *American Journal of Sociology,* 1980, *85,* 288–309.

Lodge, M., and Wahlke, J. "Psychophysiological Measures of Political Attitudes and Behavior." *Midwest Journal of Political Science,* 1972, *16,* 505–537.

Mackenzie, W. *Biological Ideas in Politics.* New York: Penguin Books, 1978.

Masters, R. D. "The Impact of Ethology on Political Science." In A. Somit (Ed.), *Biology and Politics.* The Hague: Mouton, 1976a.

Masters, R. D. "Functional Approaches to Analogical Comparison Between Species." In M. von Cranach (Ed.), *Methods of Inference from Animal to Human Behavior.* The Hague: Mouton, 1976b.

Peterson, S. A., and Somit, A. "Methodological Problems Associated with a Biologically Oriented Social Science." *Journal of Social Biological Structures,* 1978, *1,* 11–25.

Pribram, K. H. "Behaviorism, Phenomenology, and Holism in Psychology: A Scientific Analysis." *Journal of Social and Biological Structures,* 1979, *2* (1), 65–72.

Rankin, R. E., and Campbell, D. T. "Galvanic Skin Response to Negro and White Experimenters." *Journal of Abnormal and Social Psychology,* 1955, *51,* 30–33.

Schubert, G. "Politics as a Life Science: How and Why the Impact of Modern Biology Will Revolutionize the Study of Political Behavior." In A. Somit (Ed.), *Biology and Politics.* The Hague: Mouton, 1976.

Schwartz, D. C. "Somatic States and Political Behavior: An Interpretation and Empirical Extension of Politics." In A. Somit (Ed.), *Biology and Politics.* The Hague: Mouton, 1976.

Schwartz, D. C. "Toward a More Relevant and Rigorous Political Science." *Journal of Politics,* 1974, *36,* 103–137.

Schwartz, D. C., Garrison, J., and Alouf, J. "Health, Body Images, and Political Socialization." In D. Schwartz and S. Schwartz (Eds.), *New Directions in Political Socialization.* New York: Free Press, 1975.

Somit, A. "Biopolitics." *British Journal of Political Science,* 1972, *2,* 209–238.

Somit, A. "Introduction." In A. Somit (Ed.), *Biology and Politics.* The Hague: Mouton, 1976.

Somit, A. "Toward a More Biologically Oriented Political Science: Ethology and Psychopharmacology." *Midwest Journal of Political Science,* 1968, *12,* 550–567.

Somit, A., Peterson, S. A., Richardson, W. D., and Goldfischer, D. S. *The Literature of Biopolitics.* (rev. ed.) De Kalb: Center for Biopolitical Research, Northern Illinois University, 1980.

Tursky, B., Lodge, M., and Cross, D. "A Bio-Behavioral Framework for the Analysis of Political Behavior." In A. Somit (Ed.), *Biology and Politics.* The Hague: Mouton, 1976.

von Cranach, M. (Ed.). *Methods of Inference from Animal to Human Behavior.* The Hague: Mouton, 1976.

Wahlke, J. "Observations on Biopolitical Study." In A. Somit (Ed.), *Biology and Politics.* The Hague: Mouton, 1976.

Watts, M. W., and Sumi, D. "Desensitization of Children to Violence? Another Look at Television's Effects." *Experimental Study of Politics,* 1976, *5,* 1–24.

Watts, M. W., and Sumi, D. "Studies in the Physiological Component of Aggression-Related Social Attitudes." *American Journal of Political Science,* 1979, *23,* 528–558.

Watts, M. W. "Psychophysiological Analysis of Personality/Attitude Scales: Some Experimental Results." *Political Methodology,* forthcoming.

Wiegele, T. C. *Biopolitics: Search for a More Human Political Science.* Boulder: Westview Press, 1979.

Willhoite, F., Jr. "Equal Opportunity and Primate Particularism." *Journal of Politics,* 1975, *37,* 270–276.

Willhoite, F., Jr. "Ethology and the Tradition of Political Thought." *Journal of Politics,* 1971, *33,* 615–641.

Willhoite, F., Jr. "Evolution and Collective Intolerance." *Journal of Politics,* 1977, *39,* 667–684.

Willhoite, F., Jr. "Primates and Political Authority: A Biobehavioral Perspective." *American Political Science Review,* 1976, *70,* 1110–1126.

Meredith W. Watts is professor of political science and assistant chancellor, University of Wisconsin–Milwaukee. He has written on political behavior and attitudes, and on biopolitics.

Social scientists must open themselves to the methodologies and theories of ethologists and comparative psychologists.

The Use of Ethological Methods in Political Analysis

Glendon Schubert

From a biological perspective, many human behaviors—both individual and social—evince widespread and important homologies and analogies with the behaviors of other animals, especially mammals, and among them, other primates, carnivores, and cetaceans. Ethological methods emphasize the importance of observing nonverbal communication, of using video and audio recordings to support detailed laboratory analysis of social communication, and of evaluating behavior in terms of evolutionary function and physiological cause as well as physical form. Most work in human ethology has focused upon either nonverbal communication or the structure of authority in groups of young children. I have discussed elsewhere (Schubert, 1979a; 1979b; 1979c; in press) the substantive trends in ethological research, including the beginning efforts of political scientists; here we shall focus upon questions of methodology per se.

At least one reason for the dearth of research in political ethology is the lack of awareness, among political scientists, both of what kinds of questions about politics are appropriate for ethological inquiry and of how to do ethological research, given such a focus of investigation. So, the justification for this chapter is the same as for any other discussion of methodology—that is, it is intended to facilitate the doing of what might be done and what needs to be done. The assumption that there is need for such guidance seems justified by the disinclination of political scientists to fraternize with zoologists or comparative psychologists who study the behavior of animals other than humans.

But surely the best way to learn ethology is to get to know some ethologists well enough to become a participant observer in their work and thereby to discover, at an operational rather than ideological level, why and how they do it. In addition to the several years that I have spent in that kind of personal experience, I have endeavored to systematize my familiarity with the relevant research literature on ethological methodology and particularly with the literature of the past decade. The publication of the first general work on the subject by a zoologist (Lehner, 1979) after the first draft of the present paper had been completed enabled me to check my own findings and conclusions against those of an experienced and quantitatively oriented field ethologist; and naturally I have placed special reliance upon, and shall make frequent reference to, Lehner's *Handbook* in the discussion that follows. For an earlier handbook focusing upon methodology in human ethological studies of child development, see Hutt and Hutt (1970); and for a symposium emphasizing the method of naturalistic observation in comparative psychology, see Willems and Rausch (1969).

The structure of this paper is as follows: First, we shall consider the epistemology of modern ethological research, with its strong emphasis upon naturalistic observation guided by evolutionary theory. Then, the habitats in which ethologists themselves typically behave and the sensory modalities of particular interest to ethologists are discussed. Next, the instruments used in ethological analysis are reviewed, and questions of research design and of statistical and computer analysis are asked. The concluding section deals with ethological study of social structure, with particular reference to the possible use in research of political ethology of the model of social organization and associated field data collection that was proposed by a committee of the International Ethological Association (Glen McBride, Appendix D, in Lehner, 1979).

Ethological Epistemology

Purton, a philosopher of science, has said: "Ethologists categorize behaviour according to its form, its function and its causation. . . . To some extent the ethologist is bound to classify behaviour simply by its physical form ('nodding', 'facing away', 'ventral roll' . . .) but this sort of classification by form is distinct from both causal classification and functional classification. . . . On the whole, ethologists first classify displays according to their formal characteristics and then begin to enquire about function. In this sense the formal system has a certain priority. On the other hand, once a reasonably detailed knowledge of a species' displays has been acquired, the formal characteristics will in some contexts be of only secondary interest. . . . [Moreover,] functional and causal categories do not always coincide, and . . . behaviour which is functionally out of place may be just what is expected causally. . . . In addition, a fourth scheme plays an important role, i.e., our ordinary scheme of talk about human actions which centres around questions concerning motive or purpose. These four frameworks are conceptually distinct; they organize the data in different ways so that what counts as 'the same' sort of behaviour may be different in each of the schemes. On the other hand, there are links between the schemes, some of which are logical, some empirical, and some theoretical" (Purton, 1978, pp. 653, 668).

Hinde (1970, chap. 2) classifies behavior by its immediate causation (for example, agonistic: response to rival male) and its functional (evolutionarily adaptive; for example, hunting, threat, courtship) consequences; he subsumes form under both types, treating it as a question of description rather than of classification. In place of physical form, Hinde proposes historical classification, and this in turn subsumes two distinct subtypes: by source or origin, as in the case of fixed action patterns; and by method of acquisition, such as learning or ritualization.

Two of the Hinde classification types, causal and historical, are juxtaposed with the minimal and maximal levels of specific size to define a fourfold table that evidently underlies the four well-known whys of animal behavior posed by Tinbergen in 1951 and summarized in Table 1. Blurton-Jones has commented (1972, p. 8): "This division into four kinds of question simplifies many features of studying behavior and avoids much confusion, particularly between proximate and ultimate causes and between features of learning (development) and features of motivation (causation). It should also prevent confusion of adaptation and evolution with development, but seldom has . . . . [and] failures to appreciate this [are] one reason for the stubborn persistence of nature-nurture arguments in research on animal behavior."

But the most difficult as well as most important classification is functional (Baerends, 1975, 1976), and this raises the question of comparative method, for reasons that Bateson (1968, p. 392) and Purton (1978, p. 668)

Table 1. The Four Tinbergen Whys of Behavior

Level of Analysis	Types of Classification	
	Causal	Historical
Individual	Physiology 1	Development 2
Species	3 Function	4 Evolution

Legend: The four questions:
1. "What made it do it now?"
2. "How did this individual grow up to be an animal who responds that way?"
3. "What use is this to the animal? Why do these animals do this sort of thing? What do they get from it?"
4. "Why does this kind of animal solve this problem of survival in this particular way?"
Source: Blurton-Jones (1972, p. 8).

have discussed and that Hinde (1974, p. 6; compare pp. 4–5) has explained as follows: "Comparative study of different species also gives us some under-standing of the biological functions of behaviour, the ways in which it has become adapted in evolution to enhance survival and reproduction. The con-cept of function depends on the study of differences between organisms, and we could make little progress if we confined our studies to one species." With more explicit reference to human behavioral functions, Blurton-Jones (1975, pp. 83–84; compare pp. 84–88) has suggested: "Where contemporary agricul-tural and industrial man is concerned, the extrapolations may be useful (as are straight comparisons that show up differences between man and any other animal) in suggesting major basic patterns to human behaviour or social orga-nization. This role is not unlike that of cross-cultural comparisons and could be of comparable importance. These, like cross-species comparisons, can draw attention to the peculiarity of features we take for granted and to the generali-ties traceable in all people. They could also suggest ways of perceiving behav-iour that step completely outside our cultural preconceptions, an aim basic to any scientific approach to behaviour."

In addition to cross-species comparison, ethological epistemology is committed to methodological individualism. Hinde (1974, pp. 13–14) cites Tinbergen as authority for the proposition that "social behaviour is not an exclusive category, but merely a label for types of behaviour involving other members of the species, and that it thus cuts across the more basic categories such as feeding, sexual, and parental behaviour." Hinde then adds his own agreement: "The attractive nature of this view is apparent as soon as one con-siders the nature of social behaviour. Generalizing broadly across species for the moment, social behaviour usually consists of an appetitive phase which is brought to an end by the proximity of one or more other individuals, and is then succeeded by some other activity." But the examples that he then offers—feeding, copulation, sleeping—do not impress me as being necessarily or even characteristically individual behaviors, at least among humans.

Clutton-Brock and Harvey (1976, p. 226) have also concluded that "[f]unctional explanations of variation in primate social behaviour have been retarded by attempts to investigate the adaptive significance of social systems instead of social relationships. Examination of the costs and benefits of social acts to individuals allows specific hypotheses to be formulated and tested."

An insistent theme in virtually all discussions by ethologists about research methods is the basic and central importance, first, of observation, and then of description, of any behavior to be studied (Fagen and Goldman, 1977). According to Tinbergen (1972, p. vii), "[W]hat is now just beginning to happen in human ethology—is reminiscent of what occurred in the later 'twen-ties and early 'thirties to the science of animal behaviour; a new type of research worker is busy building the foundations of a science, by returning, with renewed attention and interest in detail, to the basic task of observation and description of the natural phenomena that have to be understood. I call this 'building' because, with due respect to human psychology in its widest sense, I consider that it is not yet really a science."

That may be one man's meat, but it is also a typical as well as authoritative exposition of the dominant viewpoint of the older generation of ethologists. As Bateson (1968, p. 392) has remarked, "The older ethologists did not always hamper themselves with precise measurements and their behavioral categories arose from their observations rather than being imposed on them." The virtues of descriptive empiricism based on "unbiased" (that is to say, atheoretical) observation and description have also been extolled by many younger ethologists, and by Blurton-Jones (1972, pp. 10-14; 1975, pp. 73-74) in particular. But description presumes a theory to focus perception and to define criteria of relevance, and, as we already have seen, virtually all ethologists accept the neo-Darwinian theory of evolution and its correlates, natural selection and functional adaptation, plus a theory of science that defines the procedures for social relations in doing research (Hull, 1978). Surely, both theories impose severe constraints in biasing the perceptions and conceptions of ethologists about the animal behaviors that they purport to observe and describe so dispassionately. Indeed, almost a century ago, such nonbiologists as Oliver Wendell Holmes, Jr., the son of a famous physician and friend of William James and of other transcendentalists, understood and wrote of the "inarticulate major premises" underlying both facts and values in the thinking of even professional judges. Perhaps Blurton-Jones and other ethological psychology skeptics might turn with profit to the research literature in legal realism, which would introduce them to fact skepticism.

This is a familiar issue to political scientists, harking back as it does to Arthur Bentley's use of the turn-of-the-century sociology of Albion-Small in arguments on behalf of a science of politics that would be a form of investigative reporting in which each precious fact would serve as its own Deep Throat. (For most of his adult life, Bentley was a political reporter for a Chicago newspaper, and his sarcastic denigration of theory as "spooks" and "ghosts" is well known.) As political scientists are well aware, Bentley's (1908) principal impact upon political science came only after four and a half decades, when *The Process of Government* was resurrected by David Truman (1951) as *The Governmental Process,* in which form it contributed to the rise of political behavioralism during the fifties. It was then Bentley's ironic fate to be recognized as the godfather of interest-group theory; the barefoot empiricism so emphatically featured in his own statement of his views on methodology persists only in the minds of a few longtime and devoted disciples. Writing a decade ago, Barnes (1968, p. 100) observed that the last Lamarckian at a British university had then only recently vacated a chair in biology; no doubt an observer of American political science hence will be able to record the superannuation of the last Bentleyan. But political scientists need no advice from ethologists about descriptive, atheoretical epistemologies of science. Indeed, political science almost certainly can enlighten ethology about that trip.

However, Blurton-Jones (1975, pp. 73-74) makes a second and, I believe, much more valuable point about method in the context of his defense of empiricism and attack upon psychology. Under the heading "Ethologists distrust interview data when they are taken to be data about behaviour," he summarizes research on attachment theory by Judith Bernal, who "reports a number of findings based on prenatal interviews with the mothers and on special diaries kept by the mothers for the ten days following the birth. We all know the difference between the way first and second babies are treated. Ask any mother. Or do we? In the prenatal interviews the mother(s) said they would not spoil the second one, they wouldn't rush to pick it up whenever it cried, they would just let it cry (this *strange* ideal our mothers strive to fit themselves to). But the diaries showed that they responded *more* quickly than to firstborns, and that the secondborns spend more time out of their cots and with the mother than the firstborns." The suggestion that attitudes should be checked against performances is hardly novel to either psychology or political science (Watts, 1979; Wahlke and Lodge, 1972; and Wahlke, 1979), but it does identify a theme common to both contemporary ethology and contemporary biopolitics.

In a paper focused upon the methodological implications of biology for political science, John Paul Scott (1978, p. 16) proposed: "From the technical viewpoint, the biological study of human behavior, and of political behavior in particular, must be largely observational rather than experimental, for obvious reasons, one of which is the nature of social control. However, observation can be quantitative and can be just as rigorous as experimental science, and it has the great advantage that it allows us to study a variety of phenomena which are not amenable to experimental control." Amplifying what is meant by *experimental*, Blurton-Jones (1972, p. 10) remarked, "It would be quite wrong to think that ethology is nonexperimental. . . . Many experiments can be done in the field, in an animal or a person's natural environment. . . . An experiment normally consists of some controlled modification of the situation where the modified and the unmodified situation are compared for their effects on the animal in order to disprove one of a pair of hypotheses. Whether these are done on animals or people or indoors or outdoors is quite irrelevant."

Similarly, Lehner (1979, pp. 59–60) distinguishes between naturalistic observation and experimental manipulation. The former involves an ethologist in collecting facts that may aid in understanding phenomena or lead to the formulation of hypotheses, while the latter requires controlling the conditions under which an event occurs and eliminating· extraneous influences as much as possible, so that close observation can better reveal relationships between the variables upon which attention is focused. Human intrusion is minimized in naturalistic observation, although, Lehner argues, for at least some phyla (for example, insects and fishes), artificially constructed laboratory environments can closely approximate natural habitats. In any event, naturalistic observation and experimental manipulation are best conceived of as complementary, with most long-term research programs involving both approaches in counterpoint. Both approaches apply to either field or laboratory work, and Lehner (1979, p. 61) quotes Menzel (1969, p. 78) to the effect that "any sharp division between naturalistic and experimental methodology is not only undesirable but impossible."

Ethologist Habitats

As a matter of fact, most of the work in classical ethology was done either in the field or else by zoo watchers, but comparative psychology came out of a tradition that was habituated to laboratory experimentation, and the modern field of ethology works with all three loci — in the field, in laboratories, and in zoos. These very different settings impose different requirements upon access to, observation of, and recording of animal behavior. It does not strain the normal uses of language to note that all three settings also apply (in principle) to human ethology; all we need to do is to substitute institutional confinement (in hospitals, prisons, schools and universities, military camps, and so forth) for the word *zoo*, which we employ for the imprisonment of nonhuman animals. Bateson (1968, pp. 389, 392, 394) has remarked, "If an animal is to be studied in an unrestricted environment and the measures of its behavior are to be quantitive, special techniques of observation and recording are needed. . . . [But then ethological observation] in turn leads [the researcher] back to the laboratory. . . . The need for quantitative evidence in observational studies raises two main problems. The first is to decide on those features of behavior that need recording. The second is to record them reliably when they occur. It has been in resolving these difficulties that the ways of the ethologist and the psychologist have unnecessarily parted. . . . [Thus, even] when the observer has decided what to look for, his problems are by no means over. He must decide on a method which will provide him with a permanent record of what he has observed."

Lehner (1979, pp. 89–107) distinguishes between natural variation and artificial manipulation, and also between experimental manipulation of the animal and of the environment, in either field or laboratory. Let us begin with natural variation, in which the observer takes advantage of changes that occur naturally in both the animal and its environment and hypothesizes that species behavior also will vary accordingly. An excellent example is provided by the work of Barash (1974) on the effect of habitat variation upon the social structures of woodchucks and marmots, animals that are solitary in Eastern woodlands but highly gregarious in Western mountain meadows. Masters (1978) has explicitly related this research to proposals for the ethological study of human political behavior. However, there is a long tradition of artificial manipulation in classical ethology, with models (dummy animals or parts thereof) being used as stimuli to release specific behaviors of the subject animals, as exemplified by Lorenz and Tinbergen's goose hawk silhouettes, by Tinbergen's use of cardboard replicas of herring gull heads on which the placement of the orange spot that nestlings peck to obtain food varies, and by Tinbergen and Baerends's use of differently sized, colored, and shaped wooden eggs to test nesting gull preferences.

When feral animals are manipulated (as by marking, including, though not limited to, that done for identification purposes, or through temporary capture — often by tranquilization — for installation of biotelemetric or other equipment or for release elsewhere), Lehner remarks that it is important to observe the effects not only upon the subject animals but also upon the con-specific and nonspecific animals with which they interact. Lehner distin-guishes three types of field environmental manipulation: intraspecific, as would be accomplished, for example, by removing the dominant male from a troop of patas monkeys; interspecific, as was unwittingly accomplished when lamprey eels a generation ago used the St. Lawrence Ship Canal to enter the Great Lakes; and manipulation of the physical environment, as would be exemplified by the building of additional dams in the Tennessee River, which would add the snail darter to the growing list of recently extinct species, or as in the relocation of an animal to a similar but different habitat.

The manipulation of laboratory animals is, of course, the name of the game for experimental psychology and for much biomedical research, but it is also illustrated by the unanticipated consequences of the insertion of an intra-uterine device in the subdominant mature female of a laboratory group of monkeys in a communal cage: The intent was to reduce population increase for this group, but the effect was to eliminate the previously more dominant female monkey, who was soon so badly wounded by her companions that the experimenters felt impelled to sacrifice her (Chance and others, 1977). Again, a principal justification for laboratory ethology is its advantages for controlling and manipulating the environment in which the behavior of subject animals can be observed. A fairly recent development in American ethological work has been the establishment of relatively large outdoor enclosures (with "enriched" if not exactly natural environments) for long-range and multifac-eted studies of social groups of conspecifics, particularly of primates and wolves. There are also several island reserves, such as Caya Santiago in the Caribbean Sea, on each of which a particular primate species was introduced some time ago; these species are now well established, and human access to these reserves is available only to qualified researchers. The main trend in eth-ological research has been in the direction of recognizing the value of both field and laboratory settings, utilizing either or both together depending upon the problems and hypotheses to be investigated (Price and Stokes, 1975).

Sensory Modalities

To be sure, ethologists are concerned with the study of animals that orient themselves by means of a broad and complex array of sensory modali-ties, with smell and taste and touch assuming relatively greater overall impor-tance in relation to sound and vision than for humans, and with additional taxes unknown among at least contemporary humans. Like all other primates, humans are exceptionally biased in favor of visual orientation, and that con-straint applies also to ethologists. But, in addition to the visual surveillance that observation literally implies, ethologists observe various animals through all the other modalities open to human sense perception. Presumably vision and audition will constitute the principal modes of observation for political ethologists, but this expectation in no way precludes the possibilities of imaginative research designs that literally feel what political actors are doing and that capitalize upon chemical cues.

At the same time, it is important, as Lehner (1979, pp. 127–136) emphasizes, for observers to be self-consciously aware of the potential for their own unwitting contributions to observational error. Field observation, in particular, typically involves a transactional relationship between the animal(s) and the ethologist, and differences in their respective specific sensory capacities often result in observer effect, with the presence of the ethologist stimulating the subject animals to behave in ways other than he intends or prefers. Either or both environmental or (his own) physiological perturbations can affect the ethologist's perceptions, thereby resulting in apprehension error. Observer error can be caused by his inexperience or inadequate training; by the dynamics of positional change through time on the part of either the observer or the subject animals; and by the accumulation of fatigue, discomfort, or boredom in the observer. Observer bias refers to the attitudes and expectations of the ethologist prior to beginning the observational task. In addition, there are recording errors and errors in coding and computation. All such errors reduce both the reliability and the validity of empirical findings based on observations. Lehner discusses standard methods of measuring and improving both intraobserver and interobserver reliability, but here we are on a firm social science footing, dealing with classic problems of methodology with which any properly trained political scientist is (or ought to be) familiar. Lehner also discusses (1979, pp. 137–146) the identification and naming of individuals; and although this has not typically been a problem in conventional social science survey or laboratory research in the past, there are many possible situations in which political ethologists using direct naturalistic observation may have to cope with the classification of the behavior of human actors for whom no ready and apparent identification is at hand. In such an event, Lehner's useful survey of ethological practice with nonhuman animals, including other primates, and his remarks about the implications for the observer's own attitudes of the assignment of individual labels (including nicknames) to subject animals, will become more relevant for political ethologists.

Ethological Instrumentation

Of course, political and other social scientists have been undertaking occasional observational research on political actors for a long time, especially of temporarily captive (that is, enclosed) groups of political animals such as committees or councils, legislative chambers, and courtrooms (Mileski, 1977); this has tended to be precisely the sort of ad hoc, inductive empiricism that Blurton-Jones idealizes in principle but eschews in practice. Even the field observational study of decision making in process by the judges of the Swiss Federal Tribunal (Schubert, in press) that I carried out in 1970 was initially designed upon a model of social science laboratory theory and methods (Robert F. Bales's interaction process analysis). It is true that an index of the physiological variable of arousal, incorporating postural and other nonverbal communicative behavior, was added to the design while the pretest pilot observation was under way, and this was in full accord with ethological thinking about such matters, both substantively and procedurally. But the project was much less ethological in concept and in practice than it might have been, if only I had then been ready to take more seriously the ethological work of which I was just becoming aware. As Bateson remarks, the choice of what to observe and of where and how (in terms of sensory modalities) to observe it ought to imply also a choice of equipment for recordation, but I arrived in Switzerland with neither any specialized equipment for visual observation nor any funds budgeted to acquire some. I did bring a portable cassette recorder (plus accessories), so that I could record the sound of the individual interviews with the judges, the principal research activity, which was ongoing together with the observational project. I did also consider taking equipment into the courtroom to facilitate the coding of visual observations, but one look at an Interaction Process Recorder, the antiquarian and overpriced mechanical contraption merchandised by Bales himself, convinced me that, however suitable it may be for use behind one-way mirrors in laboratory settings, no American encumbered with that kind of obstrusive gimmick would last for two minutes in the audience of a panel of Swiss judges who more or less resented his presence in any case. As Bateson (1968, p. 395) has well said, "The equipment may upset the animals in a way that is not easily overcome." Someone faced with a similar problem today could use such a device as the Datamyte 900 compact, a portable electronic data collector with keyboard and computer interface that is about the size, shape, and weight of a two-ring looseleaf notebook; this instrument is described below.

An ethologist who is going to spend a few freezing hours or days in an elevated canvas hide to make a videotape with synchronized sound of, say, the nesting behavior of a colony of herring gulls on polder flats barely half a mile from an estuary of the North Sea in the early springtime will take with him a handcart full of equipment worth several thousand dollars. He does this because ethologists have learned how much information is lost or distorted when even two or three persons with binoculars—the maximum possible in such a hide—are packed in to attempt to coordinate their viewing of more than a hundred animals in continuing activity in a four-dimensional space. The taped record permits them to focus on a particular action sequence, slowing it down or running it backward or repeating it, and to carry out much more sophisticated secondary analyses of the audio data than is conceivable for a human ear confronted with several hours of the live cacaphony that typically is produced by the vocalizations of a large number of contiguous gulls. Again, however, Bateson (1968, p. 395) has cautioned, "Quite apart from the expense, the time involved in examining the film or tape after the event may outweigh any additional information obtained." This is a caveat that any political scientist with even cursory experience in the content analysis of written transcriptions of human verbal behavior feels bound to treat with respect.

The answer, of course, lies in a sampling design that implies an overall plan for quantification, including the choice of statistical measures and computer software. Here we are on much more familiar ground, since we are dealing with a subject concerning which the average political scientist today is at least as sophisticated as the average ethologist. Nevertheless, the average ethologist is much more sophisticated about the use of observational instruments than the average political scientist, so it is appropriate that we should become aware of Lehner's discussion of instrumentation. At the same time, we can deal much more summarily with the more general questions of research design and presentation and of statistical and computer methods to which he devotes four chapters, a full third of his book.

A basic tool of classical ethology, and one that remains of central importance to field ethologists working with birds and mammals, consists of binoculars and spotting scopes, regarding which Lehner provides useful information and advice, both behavioral and technical. He follows this with an entire chapter on data collection equipment. In this, first (compare Sackett, 1978) he deals with data forms, their modification according to sampling techniques, and devices such as clocks and counters and more complex event recorders, including computer-compatible data loggers. The latter include the Datamyte (Scott, Torgerson, and Masi, 1977), mentioned above, and several alternative systems (Gass, 1977), with which it is compared. Lehner then discusses the use of audio tape recorders and ultrasonic detectors; sound spectographs (for similar studies in biopolitics, see the work on voice stress by Wiegele, 1977, 1978); the uses of still photography and motion pictography, film editing, and videotape; stopwatches and metronomes; and biotelemetry. Much of this information is bound to be new to most political scientists, and it should be of considerable value in the development of direct observational approaches to the analysis of political behavior.

Political scientists presumably ought not to require the detailed instruction that Lehner provides on the general methodological problems of how to design research and present its findings, how and why to use statistical tests and methods, and how to take advantage of computer facilities in analyses of appropriately quantified and coded data. Political ethologists may, nevertheless, be interested in checking the state of the art of such matters in the general study of animal behavior, and for such purposes Lehner provides a competent and detailed discussion in chapters 5, 10, 11, and 12. Also available are two recent symposia on quantitative ethology (Colgan, 1978; Hazlett, 1978) and many more specialized discussions such as those of research design (Dunbar, 1976), sampling methods (Altmann, 1974), statistics (Denenberg, 1976), and computer software (Stricklin and others, 1977).

Social Structure

Lehner points out (1979, p. 137): "Early naturalists talked about the behavior of species. We now know that there are often major differences between populations, social units, and individuals. Hence, ethological research is building from knowledge gained from studies of various individuals and groups and attempting to produce limited generalizations." As with contemporary ethologists, so also is it with political behavioralists and therefore also with political ethologists. Much classical ethology, however, was concerned with the description and analysis of animal displays that were imputed to be specific and under almost completely genetic control. Such displays typically are presumed to have evolved and to function in order to convey primarily interspecific information in social groups or animals, and Lehner mentions several examples (pp. 206–212) of the quantified analyses of social communication that ethologists have made concerning various kinds of animals. Notwithstanding the caveats that I have expressed in another context (Schubert, 1979b), it seems probable that much human communication, especially nonverbal communication, is under considerably greater genetic control than the heavily environmentally biased social science of the past half-century has been willing—indeed, able—to concede (Eibl-Eisenfeldt, 1974; Ekman and Friesen, 1975; Hooff, 1972; and Andrews, 1963). However, except for my own pilot study of almost a decade ago (Schubert, in press), little seems to have been done in observational research on nonverbal political communication.

Lehner discusses (pp. 212–219) dominance hierarchies as a second facet of animal social behavior, and with a focus upon preschool children, this has been a major emphasis in recent work in human ethology (Barner-Barry, 1977, 1978; Omark, Strayer, and Freedman, 1979). But the simple linear relationships initially reported half a century ago by Schjelderup-Ebbe (1922) for domestic hens and a generation ago in some of the early reports of field primatology (DeVore, 1965) have been displaced by more recent work (for example, Chase, 1974), which recognizes that, in the social groups of many species, there are different hierarchies for different purposes; that hierarchies are developmental and dynamic structures; and that intransitivities are not necessarily either mistakes or unstable—other animals, as well as humans, have to live with continuing and unresolved ambiguities in their social relationships. Now that primatologists have begun to explore the processes of coalition formation (Waal, 1978), they are entering a territory that some political scientists have been exploring for almost a generation (Riker, 1962), which suggests the possibility that academic interchange across the two disciplines may indeed not be restricted to the one-way street perceived by most zoologists and experimental psychologists—a turn of events that I anticipated some years ago (Schubert, 1973).

The third aspect of social behavior described by Lehner (pp. 220–225) is social organization, and here he distinguishes five types of intraspecific animal groups: kin, comprising more or less closely related individuals; mating, including pairs, harems, leks, and spawning groups; colonial, formed by nesting pairs of one-male harems, with provisioning of young at the nest; survival, "formed by aggregation of randomly related, usually nonbreeding individuals who are mutually attracted by each other"; and aggregations, coincidental groups "formed by physical factors acting on migrating or moving animals" or "by attraction to a common resource, such as food or water." Human organization includes all five of these types, plus many more. It is understandable, therefore, that Lehner concludes (p. 221), "Analysis of social organizations is the most complex endeavor an ethologist can undertake, because it necessitates the integrative analysis of social behavior both within and between group members." This suggests that the highest level of ethological analysis of social behavior constitutes the baseline for inquiry for most of the social sciences. However, the fact that most of the social sciences have long been concerned with human social organization at levels more complex that those attained by other animals does not mean that social scientists have nothing to learn from ethological methods of studying social organizations.

McBride's appendix in Lehner's work (pp. 371–393) presents what he describes as a tentative model of the organization of animal societies, on the basis of which nine subsets of questions to guide data collection, totalling almost three hundred items, are suggested. The model consists of assumptions in the form of generalized statements about animal social behavior, including the following:

> Ecologically, a species is seen as a unit or subsystem within a community, harvesting energy and resources of specific types. The individuals are organized physiologically for the conversions of the chemical materials they harvest; they are equally well organized behaviourally to harvest and also to distribute the particular resources available among the individuals of their species, socially. Animals live in that part of their species' niche determined by the social arrangements they have made with their neighbours. The matrix or net of such social arrangements is the society. . . . [And] on analysis, all societies appear to combine variations in only four structural features . . . :
>
> 1. Species organize differently for periods of time, into *social phases,* for example, in the breeding and the nonbreeding seasons;
> 2. There are usually several types of individuals, or *castes,* organized organically. Examples are infants, juveniles, males, and females;
> 3. Animals normally aggregate into *groups* characteristic of the species, or they remain solitary;
> 4. The groups or individuals *disperse* in space in some regular pattern, such as territories or overlapping home ranges.
>
> Animals make behaviour, which is basically movements. Their sensory systems provide for the orientation of these movements and their adjustment by the surroundings. The movements are ordered into sequences which perform complex displays. . . . These movements or groups of movements, along with various attention-drawing

or attention-hindering features of colour and form, are the simple behavioural tools; they are assembled by animals into behavioural sentences of various lengths and complexity. . . .

It is the regularity and predictability of the adjustments between individuals comprising social relationships which gives them their stability, and thus stable characteristics to societies in time, in space, and in the organization of the behaviour of individuals . . . .

It is within the matrix of social relationships that animals organize their behaviour toward the other components of their environments, so that every action is carried out within a social context . . . .

Societies in this view are not fixed structures, but fixed systems. All levels of organization, and all of the control systems are the products of evolution by natural selection. Yet all of the societal information is carried only in the organization of the individuals, as their sensory system, their behavioural system or repertoire, and their summary or learning system. . . . An account of the society of a species should aim to examine the processes of maintenance as well as change, at every level of organization, behavioural, interactional, relationship, group, and society" [Lehner, 1979, pp. 372–376].

By no means all items in the entailed questionnaires, but at least some of the items, would direct inquiry along lines novel for most social scientists, or else they would suggest novel insights about well-established lines of inquiry, as may be suggested by the following:

1. Social Organization in Time
    1.4 Is the phase change uniform over wide areas, or are there different phases in different areas (for example, nonsynchronized breeding)?
2. Specialized Types of Animals
    2.5 Are there differences in camouflage beween castes?
3. Group Structures
    3.26 What locating behaviours are used to maintain contact between group members (for example, tail flicks or flashes, regular calls, body orientation so that nearest neighbours are visible, and so forth)?
4. The Organization of Spacing Territories
    4.18 What is the nature of the boundaries? Recognized equilibrium positions resulting from decreasing defence with increasing distance from home site? Or is defence equally intense over whole area?
    Home Ranges
    4.54 Do the individuals or groups ever aggregate (for example, at food or at roosting sites or for mobbing)?

5. Spacing Within Groups
   5.18 Do the parents influence the status of their young? Describe the influences on the interaction by which the dominance-subordinate relationships are formed.
6. Sexual Behaviour
   6.14 Describe bond-servicing interactions between the mates (for example, allogrooming, allofeeding, and so forth). Comment on their evolutionary origins (for example, derived from prebond courtship, copulations, from parent-offspring interactions).
7. Parent-Offspring Behaviour
   7.24 Describe the interactions involved in parental care (for example, body care, defence—including distraction displays—housekeeping at nest, supplying food to young—by bringing it, or leading the young to food, introducing them to food, feeding from bodily secretions or tropholaxis—supply of warmth, removal of faeces or urine, retrieving young to self or to nest, carrying young).
8. Ecological and Other Behaviour
   8.64 What resources may limit numbers? at which season? under what conditions? [Lehner, 1979, pp. 376–393]

Conclusion

Ethology emphasizes the fundamental importance of observing animals in their natural habitats and describing what they do as the basis for subsequent physiological, developmental, functional, and evolutionary analyses of the behavior of individuals and social groups of particular species. Hypotheses inferred from descriptive observations during the period of classical ethology and increasingly also from deductions from evolutionary theory stated with mathematical rigor are tested empirically with both field and laboratory data and with manipulation of both the subject animals and their environment. Considerable ingenuity has been demonstrated by ethologists who have capitalized upon natural variation in habitats, thereby making possible what in effect are natural experiments with the subject feral animals. Nevertheless, most field observational work entails more or less contamination by the observer of either the animals, their environment, or both, and methodological sophistication requires that these and other sources of error variance be themselves measured to the extent possible and included in analyses of the experimental data. Laboratory research often makes possible somewhat greater control over many such sources of error variance, but only at the cost of more or less unnatural physiological, social, and habitat stimulation of the animals, with consequent distortions of their behavior.

Ethologists are accustomed to working with animals whose sensory modalities often differ considerably from the overweening dependence upon visual and auditory perception and communication that characterizes virtually all social science research on human behavior, and sensitivity to non-human taxes will make political ethology cope with a much more complex scope of human behavior than has been recognized heretofore. Ethologists are also accustomed to working with a much more extensive and sophisticated array of observational (including auditional) instruments and other technical equipment for purposes of both field and laboratory studies of animal behavior than political scientists are familiar with in their studies of political behavior.

The increasing reliance of ethologists upon quantitative sophistication in research design and in statistical methods and computer software is paralleled by equivalent developments within political science. Similarly, ethological analyses of social structure necessarily have tended to deal with organization much less complex than that usual among humans. The difference in assumptions and perspective that interspecific comparison requires should help political ethology to become aware of and to investigate diverse aspects of human social organization that are commonly studied in animal behavior but that lie beyond the conceived bounds of relevance for contemporary social, including political, science.

References

Altmann, J. "Observational Study of Behavior: Sampling Methods." *Behaviour,* 1974, *49* (3), 227–265.

Andrews, R. J. "The Origin and Evolution of the Calls and Facial Expressions of the Primates." *Behaviour,* 1963, *20,* 1–109.

Baerends, G. "An Evaluation of the Conflict Hypothesis as an Explanatory Principle for the Evolution of Displays." In G. Baerends, C. Beer, and A. Manning (Eds.), *Function and Evolution in Behaviour: Essays in Honour of Professor Niko Tinbergen.* Oxford: Clarendon Press, 1975.

Baerends, G. "The Functional Organization of Behaviour." *Animal Behaviour,* 1976, *24,* 726–738.

Barash, D. P. "The Evolution of Marmot Societies: A General Theory." *Science,* 1974, *185,* 415–420.

Barner-Barry, C. "An Observational Study of Authority in a Preschool Peer Group." *Political Methodology,* 1977, *4,* 415–449.

Barner-Barry, C. "The Biological Correlates of Power and Authority: Dominance and Attention Structure." Paper presented at annual meeting of the American Political Science Association, New York, September 1, 1978.

Barnes, S. B. "Paradigms—Scientific and Social." *Man,* 1968, *4,* 94–102.

Bateson, P. P. G. "Ethological Methods of Observing Behavior." In L. Weiskrantz (Ed.), *Analysis of Behavioral Change.* New York: Harper & Row, 1968.

Bentley, A. F. *The Process of Government: A Study of Social Pressures.* Bloomington, Ind.: Principia Press, 1908.

Blurton-Jones, N. G. "Characteristics of Ethological Studies of Human Behaviour." In N. G. Blurton-Jones (Ed.), *Ethological Studies of Child Behaviour.* Cambridge: Cambridge University Press, 1972.

Blurton-Jones, N. G. "Ethology, Anthropology, and Childhood." In R. Fox (Ed.), *Biosocial Anthropology*. New York: Wiley, 1975.

Chance, M. R. A., Emory, G., and Payne, R. "Status Referents in Long-Tailed Macaques (*Macaca fascicularis*): Precursors and Effects of a Female Rebellion." *Primates,* 1977, *18,* 611–632.

Chase, I. D. "Models of Hierarchy Formation in Animal Societies." *Behavioral Science,* 1974, *19,* 374–382.

Clutton-Brock, T. H., and Harvey, P. H. "Evolutionary Rules and Primate Societies." In P. P. G. Bateson and R. A. Hinde (Eds.), *Growing Points in Ethology.* New York: Cambridge University Press, 1976.

Colgan, P. (Ed.). *Quantitative Ethology.* New York: Wiley, 1978.

Conger, R. D., and McLeod, D. "Describing Behavior in Small Groups with the Datamyte Event Recorder." *Behavioral Research Methods Instrumentation,* 1977, *9* (5), 418–424.

Denenberg, V. H. *Statistics and Experimental Design for Behavioral and Biological Researchers.* Washington, D.C.: Hemisphere, 1976.

DeVore, B. I. (Ed.). *Primate Behavior: Field Studies of Monkeys and Apes.* New York: Holt, Rinehart and Winston, 1965.

Dunbar, R. I. M. "Some Aspects of Research Design and Their Implications in the Observational Study of Behaviour." *Behaviour,* 1976, *58* (1–2), 78–98.

Eibl-Eisenfeldt, I. *Love and Hate: The Natural History of Behavior Patterns.* New York: Schocken Books, 1974. (Originally published 1970.)

Ekman, P., and Friesen, W. V. *Unmasking the Face.* Englewood Cliffs, N.J.: Prentice-Hall, 1975.

Fagen, R. M., and Goldman, R. N. "Behavioural Catalogue Analysis Methods." *Animal Behaviour,* 1977, *25,* 261–274.

Gass, C. L. "A Digital Encoder for Field Recording of Behavioral, Temporal, and Spatial Information in Directly Computer-Accessible Form." *Behavioral Research Methods and Instrumentation,* 1977, *9* (1), 5–11.

Hazlett, B. (Ed.). *Quantitative Methods in the Study of Animal Behavior.* New York: Academic Press, 1978.

Hinde, R. A. *Animal Behaviour: A Synthesis of Ethology and Comparative Psychology.* (2nd. ed.) New York: McGraw-Hill, 1970.

Hinde, R. A. *Biological Bases of Human Social Behaviour.* New York: McGraw-Hill, 1974.

Hooff, J. A. R. A. M. van. "A Comparative Approach to the Phylogeny of Laughter and Smiling." In R. A. Hinde (Ed.), *Non-Verbal Communication.* New York: Cambridge University Press, 1972.

Hull, D. L. "Altruism in Science: A Sociobiological Model of Cooperative Behaviour Among Scientists." *Animal Behaviour,* 1978, *26* (3), 685–697.

Hutt, S. J., and Hutt, C. *Direct Observations and Measurement of Behavior.* Springfield, Ill.: Thomas, 1970.

Lehner, P. *Handbook of Ethological Methods.* New York: Garland, 1979.

Masters, R. D. "Of Marmots and Men." In L. Wispe (Ed.), *Altruism, Sympathy, and Helping.* New York: Academic Press, 1978.

Menzel, E. W., Jr. "Naturalistic and Experimental Approaches to Primate Behavior." In E. P. Willems and H. L. Rausch (Eds.), *Naturalistic Viewpoints in Psychological Research.* New York: Holt, Rinehart and Winston, 1969.

Mileski, M. "Courtroom Encounters: Observation Study of a Criminal Court." *Law and Society Review,* 1977, *5* (4), 473–538.

Omark, D., Strayer, F. F., and Freedman, D. G. (Eds.). *Dominance Relations: An Ethological View of Human Social Interactions.* New York: Garland Press, 1979.

Price, E. O., and Stokes, A. W. (Eds.). *Animal Behavior in Laboratory and Field.* (2nd ed.) San Francisco: W. H. Freeman, 1975.

Purton, A. C. "Ethological Categories of Behaviour and Some Consequences of Their Conflation." *Animal Behaviour,* 1978, *6* (3), 653–670.

Riker, W. *Theory of Political Coalitions.* New Haven: Yale University Press, 1962.

Sackett, G. P. (Ed.). *Observing Behavior.* Vol. 2: *Data Collection and Analysis Methods.* Baltimore: University Park Press, 1978.

Schjelderup-Ebbe, T. "Beiträge zur Sozialpsychologie des Haushuhns." *Zeitschrift für Psychologie,* 1922, *88* (3–5), 225–252.

Schubert, G. "Biopolitical Behavior: The Nature of the Political Animal." *Polity,* 1973, *6,* 240–275.

Schubert, G. "Ethology: A Primer for Political Scientists; Part 1." *Center for Biopolitical Research Notes,* 1979a, *2* (2).

Schubert, G. "Ethology: A Primer for Political Scientists; Part 2." *Center for Biopolitical Research Notes,* 1979b, *2* (3).

Schubert, G. "Classical Ethology: Concepts and Implications for Human Ethology." *Behavioral and Brain Sciences,* 1979c, *2* (1), 44–46.

Schubert, G. "Nonverbal Communication as Political Behavior." In M. R. Key and D. Preziosi (Eds.), *Nonverbal Communication Today: Current Research.* (In press.)

Schubert, G. "The Use of Ethology in Political Analysis." (In press.)

Scott, J. P. "What Are the Expectations Regarding the Scope and Limits of Exploring the Biological Aspects of Political Behavior? Considerations of Methodology." Paper presented at annual meeting of the American Political Science Association, New York City, September 1, 1978.

Scott, K. G., Torgenson, W., and Masi, W. S. "Use of the Datamyte in Analyzing Duration of Infant Visual Behaviors." *Behavioral Research Methods and Instrumentation,* 1977, *9* (5), 429–433.

Stricklin, W. R., Graves, H. B., and Wilson, L. L. "DISTANGLE: A Fortran Program to Analyze and Simulate Spacing Behavior of Animals." *Behavioral Research Methods and Instrumentation,* 1977, *9* (4), 367–370.

Tinbergen, N. "Foreword." In N. G. Blurton-Jones (Ed.), *Ethological Studies of Child Behaviour.* Cambridge: Cambridge University Press, 1972.

Torgerson, L. "Datamyte 900." *Behavioral Research Methods and Instrumentation,* 1977, *9* (5), 405–406.

Truman, D. B. *The Governmental Process.* New York: Knopf, 1951.

Waal, F. B. M. de. "Exploitative and Familiarity-Dependent Support Strategies in a Colony of Semi-Free Living Chimpanzees." *Behaviour,* 1978, *66* (3–4), 268–312.

Wahlke, J. C. "Pre-Behavioralism in Political Science." *American Political Science Review,* 1979, *73* (1), 9–31.

Wahlke, J. C., and Lodge, M. "Psychological Measures of Political Attitudes and Behavior." *Midwest Journal of Political Science,* 1972, *16,* 505–537.

Watts, M. W. "Studies in the Physiological Component of Aggression-Related Attitudes." *American Journal of Political Science,* 1979, *23* (3), 528–558.

Wiegele, T. C. "Models of Stress and Disturbances in Elite Political Behaviors: Psychological Variables and Political Decision Making." In R. S. Robins (Ed.), *Psychopathology and Political Leadership.* Tulane Studies in Political Science, No. 16. New Orleans: Tulane University, 1977.

Wiegele, T. C. "The Psychopathology of Elite Stress in Five International Crises: A Preliminary Test of a Voice Measurement Technique." *International Studies Quarterly,* 1978, *22* (4), 467–511.

Willems, E. P., and Rausch, H. L. (Eds.). *Naturalistic Viewpoints in Psychological Research.* New York: Holt, Rinehart and Winston, 1969.

Glendon Schubert is professor of political science at the University of Hawaii at Manoa. Widely known for his empirical work on the U.S. Supreme Court, he has devoted much of the last decade to the study of ethology and biopolitics.

Comparative analysis of young children and macaque monkeys provides
a provocative view of the possibilities of the ethological approach.

The Organization and Coordination of Asymmetrical Relations Among Young Children: A Biological View of Social Power

F. F. Strayer

Social scientists interested in interdisciplinary analysis of the biological bases of human social behavior have been attracted more by the concept of evolutionary adaptation in modern behavioral biology than by the reductionist views of contemporary behavioral genetics. The awesome complexities of human social systems raise difficult and, perhaps, irresolvable problems for researchers seeking genetic explanations of human social activity. It is not surprising that, for most social scientists, the principle of genetic determinism does not seem useful as a potential conceptual link between human biology

A number of people deserve thanks for their help and collaboration with the various phases of this work. Dianne Cox and Meryl Elman helped gather the basic observational data; Roger Gauthier served as a major touchstone and springboard for the elaboration of many central ideas; Teresa Blicharski and Pierrette Precourt lent an important hand in the home stretch when deadlines were past; and Meredith Watts and the study of human social behavior. In contrast, the principles of individual adaptation and collective survival represent basic conceptual tools that are already widely used within both scientific disciplines. These two concepts seem likely to provide the necessary conceptual foundation for the elaboration of a viable interdisciplinary link between the biological and social sciences.

Unfortunately, social scientists committed to the development of such an interdisciplinary approach to human behavior are often shocked and frustrated by the rampant speculation that characterizes the growing popular literature on comparative behavioral evolution (Ardrey, 1966, 1970; Eibl-Eibesfeldt, 1971; Lorenz, 1967; Morgan, 1972; Morris, 1967, 1970). These popular authors, who often have already established a remarkable reputation as specialists in their particular branch of biology, appear to be brash pioneers invading the well-occupied niche of social scientists with careless indifference for existing traditions, important conceptual distinctions, and established empirical facts. Even the more scholarly syntheses of behavioral evolution in modern sociobiology (Barash, 1977; Wilson, 1975) fail to specify the diverse types of empirical information that must be obtained in order to substantiate a comparative biological analysis of human social behavior. Many of the sociobiological interpretations of human activity seem precariously similar to early theoretical views on human instincts. Analytic concepts are often employed in an abstract and formal manner that facilitates comprehension of the theoretical arguments, but avoids the most important scientific problem, namely, that of specifying precise empirical phenomena that can verify or falsify the proposed theoretical view. The development of an adequate empirical basis for the comparative analysis of social evolution is, perhaps, the most important current problem in modern behavioral biology. This problem demands the elaboration of adequate research methods as well as the refinement of both descriptive and analytic concepts. The importance and value of adequate methodology and precise concepts in comparative research has been well documented by recent advances in the biological analysis of social power within animal groups.

Biological Approaches to Social Power

The comparative biological analysis of power structures within stable social groups has traditionally entailed the systematic comparison of individual differences in the ability to control and favorably terminate dyadic episodes of social conflict. Within each dyadic context, the less dominant individ-

maintained a constant but effective editorial pressure, which brought the enterprise to completion.

I also acknowledge the financial support of the Conseil de Recherche en Sciences Humaines du Canada and of the Ministère de l'Éducation du Québec. Finally, a special thanks for the continuing collaboration of children, staff, and parents at La Garderie La Souritèque, who make our empirical work possible.

ual consistently produces gestures of appeasement or submission as predictable reactions to the other's aggressive behavior. Theoretically, the existence
of stable dominance relations and a stable power structure within the social
unit benefits all group members, because it reduces the absolute level of social
aggression by permitting the ritualized or symbolic expression of individual
differences in relative social power. The critical idea here is that the formation
of dominance relations may involve quite severe forms of physical attack,
while their maintenance merely requires mild forms of social threat that effectively communicate established differences in social status. From this perspective, a group power structure represents a biologically adapted network of
social relations that prescribes relative roles of assertion and acquiescence during conflict and thus eliminates potentially dangerous intragroup aggression
that could ultimately reduce the reproductive fitness of some group members.

The theoretical importance of social dominance as a factor in the organization of nonhuman primate societies is demonstrated most clearly by the
large number of other social processes to which it has been directly related.
Social dominance has been an important theoretical construct in the explanation of group defense (Jolly, 1972; Kummer, 1971), reactions to strange conspecifics (Jay, 1965; Ripley, 1967), affiliative preferences within the group
(Sade, 1972), social attention patterns (Chance, 1976; Emory, 1976; Pitcairn,
1976), social learning (Dolhinow and Bishop, 1972; Strayer, 1976), cultural
transmission of novel behavioral adaptations (Kawai, 1965; Tsumori, 1967),
the fission and fusion of social units (Furuya, 1960; Sugiyama, 1965), the
maintenance of kinship structures (Sade, 1965), and the ultimate reproductive
success of individual group members (Hausfater, 1975; Wilson, 1975). Perhaps because of its pervasive significance as an explanatory concept, empirical
research on social dominance among primates has had a long but often controversial history (Bernstein, 1970). Undoubtedly, a careful review of specific
problems related to the analysis of social power among nonhuman primates
should help to illustrate potential uses and abuses of social dominance as an
analytic concept in biologically oriented investigations of human social behavior.

Most of the historical controversy surrounding the study of primate
dominance has involved the fundamental problem of adequately specifying
precise behavioral referents for this abstract social concept. Among early primatologists, social dominance was often employed in a global fashion to refer to
a variety of quite different phenomena. For example, these researchers employed
dominance to describe classes of individual activity (dominance gestures),
classes of social outcomes (dominance incentives), aspects of individual personality (dominance traits), forms of social interaction (dominance struggles),
and social positions within the stable group (dominance roles or dominance
status). Such diversity in the use of dominance as an analytic term reflects a
serious conceptual confusion about necessary analytic distinctions in the empirical analysis of social behavior and group structure.

Bernstein (1980) argues that much of this historical confusion in research
on primate social dominance resulted from the failure to distinguish ade-

quately between observable activity and the presumed causal or motivational basis for such behavior. Early primatologists usually defined social dominance operationally as the ability to maintain priority of access to specific incentives (food, water, receptive females, and so forth). However, such operationalized approaches to the evaluation of social dominance often led to unstable empirical results, which seemed to vary directly as a function of the specific incentive used in the standardized assessment. More importantly, such results seldom corresponded to observed differences in social power within the group setting (Maslow, 1936). Given the general lack of convergence among operationally standardized measures, many prominent reseachers (Bernstein and Sharpe, 1965; Hall and DeVore, 1965) suggested that the concept of dominance should be abandoned completely, because its meaning was usually too ambiguous and its significance had been exaggerated.

However, in the last fifteen years, considerable progress has been made in the clarification and refinement of social dominance as an analytic concept in primate research. Currently, dominance is used not to summarize enduring individual traits but to describe stable patterns of asymmetrical social exchange between members of a primate group. By limiting the forms of behavior used in the evaluation of dominance to naturally occurring agonism, the modern approach stresses the resolution of observed social conflict as the only direct behavioral measure of relative social power. Perhaps more importantly, this current use of dominance implies that differences in social power are not necessarily measurable by standardized assessments based upon induced competition (Smith, Harris, and Strayer, 1977). Instead, dominance must be evaluated in terms of differential behavioral roles that characterize stable interaction patterns between two individuals during the resolution of dyadic social conflict within the group context. Although this refinement in the conceptual use of social dominance severely limits the types of activity that are pertinent to the evaluation of social power among nonhuman primates, the more restricted definition offers a direct means of verifying potential relations between dominance status and other forms of social functioning that have been suggested in sociobiological theories of primate behavior (Chance and Larsen, 1976; Wilson, 1975). Questions about how dominance is influenced by individual differences such as age, sex, size, aggressiveness, or intelligence and how dominance determines other social characteristics such as reproductive fitness, leadership roles, priority of access, or social defense require empirical answers. However, neither these questions nor their ultimate answers should alter the basic behavioral referents that are associated with dominance as a descriptive analytic concept in the biological study of primate behavior.

Ethological Research on Human Dominance

During the past ten years, ethological researchers have gradually extended the methods and analytic concepts from primatology to the empirical study of social power relations among young children. These studies have been quite diversified both in their use of social dominance as a descriptive term and in their selection of research methods for the analysis of power differences. The diversity of research approaches has an important heuristic value during the early phases of exploratory research with any new species. In addition, it provides a broader empirical view of the potential scope and possible limits of traditional biological concepts, and it helps to assure that initial results are not merely procedural or definitional artifacts.

Conceptual Approaches to the Definition of Dominance. Three quite different approaches to the definition of social dominance are currently evident in human ethology research. These approaches parallel conceptual orientations that have characterized dominance research among nonhuman primates. The first views the analysis of social power as primarily a problem of developing an adequate operational definition for the evaluation of differential dominance roles within children's peer groups. In primate research, standardized competitive contests provide the best illustration of this approach. The researcher operationally defines dominance as the ability to gain access to a selected incentive and then evaluates both social relations and group structure with this objective measure. The sociometric approach to human dominance developed by Omark and Edelman (1976) uses the child's verbal report as a basic measure of his relationship with other group members. The agreement between two individuals concerning their relative "toughness" is operationally defined as an established dyadic relationship. Subsequent analysis of all established dyads permits the derivation of a social structure that reflects relative status among group members.

The second approach to dominance assessment places less emphasis upon the derivation of a single operational definition; instead, it attempts to identify a set of individual behavioral patterns that provide an empirical and ecologically valid basis for the evaluation of differences in social status. This research strategy involves demonstrating that certain behaviors in the social repertoire share imporant functional characteristics, and that they can thus be regrouped into more comprehensive descriptive categories. Human ethologists differ in the final set of individual behaviors used as a basis for the description of individual dominance gestures. The major problem here is the specification of individual social effects as necessary and sufficient for the identification of activities that must be included in the category of dominance behaviors. For example, Chance (1976) suggests that activities which attract social attention are often expressions of social dominance. Attracting attention seems to be the most important function of what have been called dominance displays—for example, locomotor postures, the elevation of a monkey's tail, or the position of a child's chin. However, it is not clear that all types of attention-getting activities necessarily express social dominance. In contrast, Rowell (1974) places primary emphasis upon appeasement or submission as the critical consequent of dominance gestures. Such a criterion substantially reduces the number of behavioral patterns that must be included in the category of dominance activities. These two examples illustrate the current problem of determining appropriate behavioral categories for the observational assessment of social dominance, and they also raise important empirical questions about the organization of social behavior within stable peer groups. Do individuals who successfully elicit submissive reactions from peers also engage in more attention-getting display activity? Similar empirical questions are implicit in suggestions that all forms of social influence or social control represent expressions of social dominance. If the behavioral basis for the assessment of social dominance includes any form of social influence, we must demonstrate that our various subcategories of behavior defined in terms of their specific social effect (that is, submission, attention, change of another's behavior, and so forth) all provide essentially convergent information about dominance relations in different groups and in different observational settings. Our present inability to select the most appropriate index from among these various possible subcategories clearly reveals the limits of empirical knowledge in current ethological research on human dominance.

The final approach to the assessment of social dominance examines specific forms of social exchange as the primary observational data necessary for the derivation of dominance relations and group status. From this perspective, dominance cannot be discussed without simultaneous reference to social submission. Although a specific behavioral activity may have a high probability of leading to submission, only when it actually produces a submissive reaction from the social partner are we justified in concluding that is has functioned as a dominance signal. The identification of dominance exchanges requires the observer to describe how patterns of behavior are coordinated between two individuals as they participate in social interaction. Descriptive listings of individual action patterns as examples of dominance gestures provide only a summary of behaviors that may potentially function to express dominance. To determine whether such behaviors have actually been used as dominance signals, we must always examine how such acts have functioned in the specific sequence of observed social exchange. One of the more important advantages of this final descriptive approach is that it forces the observer to pay more attention to the ways in which behavior patterns are organized within a social sequence. Such emphasis upon the temporal and behavioral context of social actions may eventually reveal important causal and functional differences among activities that are difficult to distinguish at a morphological level (for example, wrestling versus rough-and-tumble play, cooperative role-play versus social control, and so forth).

Analytic Approaches to the Evaluation of Dominance. Two quite different analytic procedures are currently employed for the derivation of group status ranking. The first approach emphasizes individual differences in the organization of personal response styles. Hinde (1974) provides one of the clearest formulations of this analytic perspective when he argues that dominance, like aggressiveness, is best understood as an intervening variable that summarizes a complex set of relations between certain antecedent conditions experienced by the individual (that is, independent variables) and characteristic patterns of response to such stimulation (that is, dependent variables). Individuals who respond to such conditions with confident and assertive reactions are assumed to have a higher social status than individuals who respond with timid and self-effacing behaviors. From this perspective, social dominance is conceptualized as a dimension of personality. Differences in dominance status reflect basically different styles of social participation that result from differences in individual adaptation in a variety of social situations.

The second analytic approach to social status places greater emphasis upon the dynamics of social interaction. Although the analysis of dominance relations requires the observer to note the differential responsiveness of two individuals during social interaction, the evaluation of dominance is not reduced to a calculus of individual differences in general behavioral traits. Furthermore, since the group status hierarchy summarizes a network of such observed dyadic relationships, it cannot be equated with simple rank orderings of group members according to differential levels of engagement in some predetermined behavioral activity.

Individual rankings on aggressiveness or submissiveness can be more or less correlated with status rank in the group dominance hierarchy, but the correspondence between dominance status and ranking according to frequency of social activity will always depend upon the dynamics that characterize the particular social group: When a majority of dominance struggles occurs between high-status individuals, it is evident that these same group members will show a higher rate of both aggressive and submissive acts. In such a case, rank orderings of group members according to their total frequency of either activity will correlate positively with higher position in the status hierarchy. However, in a hypothetical second group where lower-status individuals engage more often in social conflict, we might find exactly the opposite relation between behavioral rankings and social status. Although this second status hierarchy may be identical to our earlier example at a structural level, it is clear that the distribution of social conflict would shift within the dyadic matrix. More importantly, the correlations between dominance status and individual rankings according to frequency of either aggressive or submissive behaviors would now be negative (Strayer, 1980).

Although these hypothetical examples reveal important potential differences between an individual and a social approach to the derivation of a group status ranking, in actual studies it is often quite difficult to determine which analytic procedure has been used. This confusion results from the fact that each procedure gives primary emphasis to describing a single rank ordering of group members. Even researchers who employ the social network analysis of dominance relations often fail to describe adequately the hierarchical organization of the obtained dominance structure. Figure 1 illustrates the analytic steps involved in the network approach. In order to evaluate the dominance structure for a group of macaque monkeys, it was first necessary to identify types of social behavior that characteristically constituted conflict interactions within the group. Next, the dyadic occurrence of such behaviors was

Figure 1. Macaque Dominance Structure

INITIATOR \ TARGET	A	B	C	D	E	F	G	H	I	J	K	L	M	N	O	P	total
♂ A		5	17	6	7	7	7	3	2	7	5	4	3	4	3	6	86
♀ B	1		13	3	4	2	4	1	1	1		6	1		1	2	40
♂ C	1			24	19	18	20	22	4	8	6	13	5	8	4	7	159
♀ D		1			4	5	9	2	2	1		2	3	6		1	36
♂ E						6	7	4	5	3	2	3	4	2	5	3	44
♀ F			2	1	1		3	3	2	8	7	11	2	2	1	7	50
♀ G				2				5	5	5	4	8	2	1	4	4	40
♂ H							1		1	4	3	2	4	2	1	1	19
♀ I										3				1		5	9
♂ J								1			2			1	2	3	9
♂ K												=	=	5	4	2	11
♀ L								1			=		=	1	3	4	9
♂ M											=	=		=	1	3	4
♂ N													=		2	3	5
♀ O																=	0
♀ P															=		0
total	2	6	32	36	35	38	51	42	22	40	29	49	24	33	31	51	521

examined for all possible pairs of animals in the group. Finally, the systematic graphic rearrangement of rows and columns in the dyadic matrix provided a simple linear ordering of individuals that summarized the network of dyadic relations within the group. This simple ordering of individuals maximized the number of observed conflict episodes, which are located in the upper half of the dyadic matrix. At the dyadic level, one member generally submits to the other. The degree of such role differentiation provides a measure of the rigidity of social relations for the group. In this particular primate group, 98 percent of all observed conflict was unidirectional; thus, the observed dominance relations were quite rigid and asymmetrical. Furthermore, at the level of social structure, dominance relations were governed by a linear transitivity rule. In all established dyadic relations, if one individual dominated another, but the other dominated a third individual, then the third never dominated the first. A measure of the exceptions to such linear transitivity provides a measure of how well the observed social relations fit the linear model of social dominance. In a case of good fit, all the observed relations correspond to predictions from the linear model. The obtained power structure for this group was 100 percent linear.

During the past four years, work at our laboratory has focused upon the systematic extension of this analytic strategy to the evaluation of power relations within groups of young children. Figure 2 shows an analogous dyadic matrix constructed for one of our preschool samples. Although substantially fewer established dyadic relations were evident here than in the monkey example, the nature of the observed dyadic relations and their organization into a linear pattern corresponded quite well with expectations based upon research with nonhuman primates. These analyses of social agonism, dyadic dominance, and group structure have been replicated both in our own laboratory and by other researchers. A number of analyses have shown that young children tend to maintain stable dyadic power relations that closely parallel those of nonhuman primates.*

However, the existence of stable dominance relations and the organization of such relations into a linear power structure in children's peer groups merely shows a structural similarity in the nature of spontaneously occurring agonistic behavior among human and nonhuman primates. More important questions about social dominance among young children involve how dominance relations and the group dominance structure relate to other activity within the natural group.

Social Dominance and the Coordination of Peer-Group Activity. Our most recent research has attempted to relate social dominance among young children to other types of social influence, which can loosely be called leadership activities, that are readily observable among young children. The central question guiding this research was the extent to which knowledge about group dominance status could provide reliable information about other forms of social influence. Our research strategy entailed making an objective assessment of social patterns for the resolution of dyadic conflict while independently evaluating instances of social influence. Two preschool groups were observed. The first consisted of fifteen children (seven girls and eight boys) who ranged in age from three to four years (mean age: forty-five months). The second group contained fourteen children (seven girls and seven boys) who ranged in age from four to nearly six years (mean age: sixty-two months). Both groups were observed during free-play periods over an eight-week observational period. Video records of social interaction were collected daily for each child using a five-minute focal individual sample. Our evaluation of social dominance required scanning each day's video records for instances of social conflict in order to obtain as much information as possible. Estimates of

*For reasons of space, several illustrative figures had to be deleted here and later in the text. The basic meaning is retained, although some graphic lucidity is lost. It can only be hoped that this discussion will provoke the reader to consult other works by the author.

Figure 2. Langara Preschool Dominance Structure

INITIATOR \ TARGET	Ro	Ss	Br	If	Td	Sd	Pe	Ir	Cs	Ka	Ch	Ty	Gl	Sa	Me	Ju	Sh	total
♀ Ro	▨	1	1		1=								1	1				5
♂ Ss		▨	1=	3	1			4	1		1		3			1		5
♂ Br	1=		▨	1	1				1	6			3	2		1		6
♂ If				▨	2	1	8	2	2	1	↑↓		4	2		2	1	5
♂ Td	1=				▨	=	3		7	2						1		4
♂ Sd					=	▨	1	1	4	2	1		↑↓	1				0
♂ Pe						1	▨	1					1=	3				6
♂ Ir								▨	3	2	1							6
♂ Cs							2	1	▨	=	=		1					4
♀ Ka									=	▨	=	9	1			1	3	4
♀ Ch				1↑↓					=	=	▨	=	5	3		1	1	1
♀ Ty											=	▨	1=	=	2			3
♂ Gl							1=				1	1=	▨	=	=	3	5	1
♂ Sa				1↑↓							1	=	=	▨	=			2
♀ Me												=	=		▨	=	=	0
♀ Ju															=	▨	=	0
♀ Sh															=	=	▨	0
total	1	2	2	5	6	4	13	9	18	7	10	11	19	12	2	9	12	142

individual behavioral activity were derived from a detailed analysis of twenty minutes of free-play interaction for each of the twenty-nine children.

The behavioral analysis of social dominance entailed the identification of particular types of conflict activity noted in our previous studies. Three major categories of social aggression and three categories of response to aggression were employed. Aggressive categories included initiated *attacks, threats,* and *struggles* for objects or position. Response categories included *submissive gestures, loss* of objects or position, and *ignore.* Only interactions that included a submissive gesture or less were used in our evaluation of social dominance. The analysis of these activities within each of the two preschool groups led to the identification of dominance structures that closely paralleled those described in previous studies. The younger children had slightly less rigid and linear dominance relations. However, the nature of the social dominance structure in both groups was well within the range of variation evident from previous studies.

Our analysis of other social activities focused upon two general classes of behavior. The first class included a constellation of activities dealing with the distribution of social attention within each group. In order to evaluate the rate at which each child served as the *focus of attention,* instantaneous scan samples were obtained from thirty records of each child. In addition, the rate of *bids for attention* was coded from the video records. Bids for attention were distinguished according to whether or not the child was successful in eliciting the attention of another. The second general class of behavior included examples of direct or indirect control over the activity of another child. The most direct form of control consisted of overt intervention by the focal child into the activities of another. Such interventions were coded as *influence* if the focal child successfully altered the ongoing activity of another. A less direct form of control involved serving as a model for another group member. Such instances of imitation were coded as *copy* when it was clear that the focal child initiated the mutual activity prior to the onset of the activity by the other group member. In instances where the temporal sequencing of joint activity could not be distinguished, the activity was coded as *join,* an example of mutual or reciprocal social influence. Inter-observer reliability in the coding of each of these categories was assessed independently. Reliability coefficients were always above 80 percent.

In each group, both attention and social control were observed more frequently than social conflict. Aggressive acts were initiated with an average rate of two per hour, while submissive gestures occurred 1.75 times per hour. In contrast, bids for attention occurred with an average rate of 7.48 per hour (4.54 for successful bids and 2.94 for unsuccessful bids). Efforts at social control had an average rate of 11.64 events per hour (4.60 for influence, 3.64 for copy, and 3.40 for join). In general, forms of social influence occurred nearly ten times more frequently than social conflict. This difference in rate of occurrence provides some preliminary insight into the relative importance of these two types of social behavior among preschool children.

In order to examine the relation between position within the group dominance structure and the rate of initiated social activities, correlations were computed separately for each group between status rank and individual scores for categories of behavior within each general class of social activity. Table 1 shows the degree of relationship between position within the group's status structure and observed rates of social behavior. In the younger group, there was no consistent relation between any of the conflict measures and status rank. In the older group, higher-ranking children showed a significant tendency to initiate more aggressive exchanges. However, while this correlation is significant, dominance rank cannot be reduced to the rate of aggressiveness, since the variance in common between these two measures is only 36 percent. It is interesting that in both groups the rate of submissive gestures was also positively correlated with dominance status. This finding reflects the fact that high-status children not only direct more aggression toward their subordinates in the dominance hierarchy but that they also receive and submit to aggression from their immediate superiors.

Table 1. Behavioral Correlates of Preschool Dominance Status

Behavioral Activity	Group A (45 months)	Group B (62 months)
Conflict		
Aggressiveness	0.29	0.60*
Receipt of Aggression	0.39	-0.01
Submissiveness	0.30	0.44
Rate of Ignoring	0.18	0.35
Attention		
Receipt of Attention	0.20	0.48*
Attracting Attention	-0.09	0.23
Being Attracted	-0.37	0.25
Bids for Attention	0.03	0.07
Receipt of Bids	-0.28	0.27
Control		
Influencing	0.03	0.51
Being Influenced	0.15	-0.04
Copying	0.35	0.34
Being Copied	0.63**	0.60*
Being Joined	0.55*	0.71**

*p < .05
**p < .01

Status rankings were not consistently related to any of the attentional measures for the younger preschool group. With the older children, however, a significant relation emerged between the rate of received attention and status rank. The relation between measures of social control and position within the group dominance hierarchies were most consistent for the two groups. More dominant children were copied and joined more often by other group members. In addition, in the older group, there was a significant tendency for high status children to intervene and influence other group members. These findings suggest that position within the group status structure is associated with a greater likelihood of engaging other forms of social influence as well with successfully resolving episodes of social conflict. However, these data do not provide sufficient information about how attentional and control relationships between children vary as a function of their relative dyadic dominance. In order to gain more direct insight into coordination of these social relations in our two groups, observational records were reanalyzed with an emphasis upon dyadic direction of each class of social exchange.

Our first question concerned the extent to which each of the three major classes of social activity was characterized by dyadic asymmetry. That is, we wanted to evaluate the relative likelihood that for a given dyad each type of social activity was generally directed by one individual toward the other. The agreement among indicators varied from 72 percent to 96 percent (mean: 85.7 percent), indicating that conflict, attention, and social control tend to be unidirectional within each dyadic context.

Given this degree of dyadic asymmetry in each type of behavior, our next analysis assessed the correspondence between both social attention and social control and the group dominance structure. These analyses required graphic mapping of the network of social relations derived from each activity on the dominance structure that had been derived for each group. Figure 3

Figure 3. Networks of Social Exchange

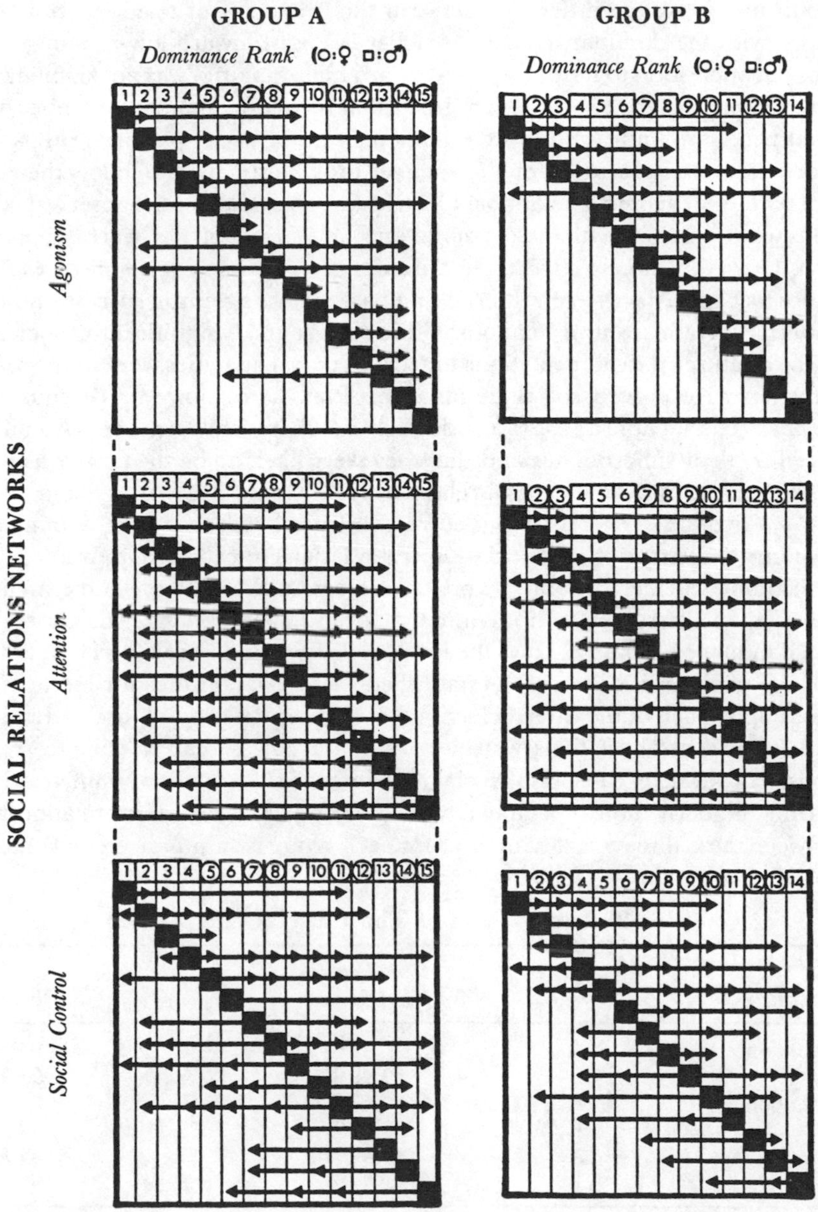

shows the resulting networks of social exchange. In this diagram, lines below the diagonal represent either nonlinear asymmetrical relations or dyadic relations in which each child was successful in half of the observed exchanges. In the agonistic network, there was a preponderance of connectedness above the diagonal, reflecting the basic linearity and rigidity of the original dominance structure. Although the network of attentional relations revealed a slight tendency for the more subordinate children to orient toward higher-ranking group members, the difference between the frequency of relations in accordance with the dominance structure (that is, high- toward low-ranking) and those in opposition to it (that is, low- toward high-ranking) was not significant. Thus, it was not the case in either group that the more dominant member of a dyad necessarily attracted more attention from the social partner. Similarly, the social control network for the younger group also failed to follow the general pattern of agonistic relations. Once again, about half of the observed relations were in accord with dyadic dominance relations, and half were in opposition. However, among children in the older group, some degree of correspondence was evident between the ranking based upon agonistic success and the pattern of dyadic control relations within the group. A significant proportion of the established dyadic relations for social control activities were in the same direction as observed social dominance. Even here, however, it must be emphasized that at least some agonistically dominant children were also influenced by their subordinates and that they were likely to be the targets as well as the leaders in dyadic control relationships.

Given the general lack of correspondence between dyadic dominance relations and attention and social control relations in each group, we decided to examine whether it would be possible to create alternate social hierarchies based upon the success of individual children in the nonagonistic classes of social influence. Table 2 shows the levels of rigidity and linearity as well as the relative number of dyads with established social relationships for hierarchies based upon each of the three behavioral categories. In general, both attention and social control relations were less rigid and linear than dominance relations. However, each class of social activity was sufficiently asymmetrical to permit the derivation of a linear rank ordering. The degree of relationship between these three social structures for each group is summarized in Table 3.

Table 2. Comparison of Three Social Hierarchies

Activity	Group	Dyads Observed	Rigidity	Linearity
Dominance	A	40%	84%	93%
	B	42%	96%	93%
Attention	A	67%	75%	82%
	B	64%	72%	90%
Control	A	42%	69%	89%
	B	47%	73%	91%

Table 3. Correlations Between Three Social Hierarchies

Group	Activity	Dominance	Attention	Control
(A)	Dominance	1.00		
	Attention	– 0.04	1.00	
	Control	. 0.33	0.02	1.00
(B)	Dominance	1.00		
	Attention	0.17	1.00	
	Control	0.76*	0.09	1.00

*$p < 0.01$

Among the younger children, it was clear that each linear structure was uncorrelated with the other two. However, among the older children, the control structure was significantly similar to the dominance hierarchy.

Discussion

The findings presented here underscore the fact that linear status structures can be derived for a number of qualitatively different forms of social exchange and that the multidimensional complexity that is intrinsic to a preschool peer group cannot be reduced to a simple, single concept such as social dominance. The increasing convergence between agonistic relations and social control roles among older children may reflect a developmental trend for the consolidation of previously independent aspects of dyadic relationships. However, such possible developmental trends require considerably more investigation. Perhaps at the most general level, the present findings stress the importance of examining the various aspects of social activity from more than a simple individual level. The significant relation between position in the group status hierarchy and amount of attention received for older children did not reflect a general disposition for subordinate children to monitor their immediate superiors. In fact, social attention relations within each group did not relate directly to dominance status. It seems that the earlier empirical research on this problem may reflect differences in social activity levels rather than differences in the development of social relationships and integration into the peer group.

With more direct reference to the study of social power, it seems appropriate to note that the forms of social influence coded in the present study occurred with a higher rate among the members of the older group. This increase in social control activity was accompanied by a reduction in the overall rate of agonistic exchanges. Perhaps the transition to forms of social influence that do not require brute force is an important stage in the development of political awareness among young children. Although such awareness may originate in the rigidly asymmetrical relations of preschool dominance, it leads eventually to a more mature sense of social influence that includes reciprocal coordination of activities toward common ends.

48

References

Ardrey, R. *The Territorial Imperative.* New York: Atheneum, 1966.

Ardrey, R. *The Social Contract.* New York: Dell, 1970.

Barash, D. P. *Sociobiology and Behavior.* New York: Elsevier North-Holland, 1977.

Bernstein, I. S. "Primate Status Hierarchies." In L. Rosenblum (Ed.), *Primate Behavior: Developments in Field and Laboratory Research.* New York: Academic Press, 1970.

Bernstein, I. S. "Dominance: A Theoretical Perspective for Ethologists." In D. Omark, F. Strayer, and D. Freeman (Eds.), *Dominance Relations.* New York: Garland Press, 1980.

Bernstein, I. S., and Sharpe, L. G. "Social Roles in a Rhesus Monkey Group." *Behaviour,* 1965, *26,* 1-2.

Chance, M. R. A. "Attention Structures as the Basis of Primate Rank Orders." In M. R. A. Chance and R. R. Larsen (Eds.), *The Social Structure of Attention.* London: John Wiley, 1976.

Chance, M. R. A., and Larsen, R. R. *The Social Structure of Attention.* London: John Wiley, 1976.

Dolhinow, P., and Bishop, H. "The Development of Motor Skills and Social Relationships among Primates." In P. Dolhinow (Ed.), *Primate Patterns.* New York: Holt, Rinehart and Winston, 1972.

Eibl-Eisenfeldt, I. *Love and Hate: On the Natural History of Basic Behavioral Patterns.* London: Methuen, 1971.

Emory, G. R. "Attention Structure as a Determinant of Social Organization in the Mandrill (*Mandrillus sphinx*) and the Gelada Baboon (*Therepithecus gelada*)." In M. R. A. Chance and R. R. Larsen (Eds.), *The Social Structure of Attention.* London: John Wiley, 1976.

Furuya, Y. "An Example of Fission of a Natural Troop of Japanese Monkeys at Gagyusan." *Primates,* 1960, *2,* 149-179.

Hall, K. R. L., and DeVore, I. "Baboon Social Behavior." In I. DeVore (Ed.), *Primate Behavior: Field Studies of Monkeys and Apes.* New York: Holt, Rinehart and Winston, 1965.

Hausfater, G. *Dominance and Reproduction in Baboons: Contributions to Primatology,* Vol. 7. Basel, Switzerland: Karger, 1975.

Hinde, R. A. *Biological Basis of Human Social Behavior.* New York: McGraw-Hill, 1974.

Jay, P. "Field Studies." In A. Schrier, H. Harlow, and F. Stollnitz (Eds.), *The Behavior of Nonhuman Primates.* Vol. 2. New York: Academic Press, 1965.

Jolly, A. *The Evolution of Primate Behavior.* New York: Macmillan, 1972.

Kawai, M. "On the System of Social Rank in a Natural Troop of Japanese Monkeys." In K. Imamishi and S. Altmann (Eds.), *Japanese Monkeys.* Chicago: Altmann, 1965.

Kummer, H. *Primate Societies: Group Techniques in Ecological Adaptation.* Chicago: University of Chicago Press, 1971.

Lorenz, K. *On Aggression.* New York: Bantam, 1967.

Maslow, A. H. "The Role of Dominance in the Social and Sexual Behavior of Infrahuman Primates: Observations at Villas Park Zoo." *Journal of Genetic Psychology,* 1936, *48,* 261-277.

Morgan, E. *The Descent of Woman.* New York: Stein and Day, 1972.

Morris, D. *The Naked Ape: A Zoologist's Study of the Human Animal.* New York: McGraw-Hill, 1967.

Morris, D. *The Human Zoo.* New York: Bantam, 1970.

Omark, D. R., and Edelman, M. S. "The Development of Attention Structures in Young Children." In M. R. A. Chance and R. R. Larsen (Eds.), *The Social Structure of Attention.* London: John Wiley, 1976.

Pitcairn, T. K. "Attention and Social Structure in *Macaca fascicularis.*" In M. R. A. Chance and R. R. Larsen (Eds.), *The Social Structure of Attention.* London: John Wiley, 1976.

Ripley, S. "Intergroup Encounters among Ceylon Gray Langurs." In S. Altmann (Ed.), *Social Communication among Primates.* Chicago: University of Chicago Press, 1967.

Rowell, T. E. "The Concept of Social Dominance." *Behavioral Biology,* 1974, *11,* 131–154.

Sade, D. "Some Aspects of Parent–Offspring and Sibling Relationship in a Group of Rhesus Monkeys with a Discussion of Grooming." *American Journal of Physical Anthropology,* 1965, *23,* 1–17.

Sade, D. "Sociometrics of *Macaca mulatta:* Linkages and Cliques in Grooming Matrices." *Folia Primatologica,* 1972, *18,* 196–223.

Smith, M. S., Harris, P. J., and Strayer, F. F. "Laboratory Methods for the Assessment of Social Dominance among Captive *Saimiri sciureus.*" *Primates,* 1977, *18,* 977–984.

Strayer, F. F. "Learning and Imitation as a Function of Social Status in Macaque Monkeys (*Macaca nemestrina*)." *Animal Behavior,* 1976, *24,* 832–848.

Strayer, F. F. "Current Problems in the Study of Social Dominance." In D. Omark, F. Strayer, and D. Freeman (Eds.), *Dominance Relations.* New York: Garland Press, 1980.

Sugiyama, Y. "On Social Change of Hanuman Langurs in their Natural Condition." *Primates,* 1965, *6,* 381–418.

Tsumori, A. "Newly Acquired Behaviors and Social Interactions of Japanese Monkeys." In S. Altmann (Ed.), *Social Communication among Primates.* Chicago: University of Chicago Press, 1967.

Wilson, E. O. *Sociobiology: The New Synthesis.* Cambridge: Belknap/Harvard University Press, 1975.

F. F. Strayer is professor of psychology and director of the Center for Interdisciplinary Research in Adaptation and Development at the University of Quebec in Montreal. He is currently responsible for the Human Ethology Laboratory at the Center and continues an active comparative research program with both human and nonhuman primates. Program activities have included investigations of social adaptation among young children as well as among chimpanzees and various species of Old and New World monkeys.

Naturalistic but controlled observation of young children can yield
considerable information about rudimentary forms of
political socialization.

Longitudinal Observational Research and the Study of Basic Forms of Political Socialization

Carol Barner-Barry

Observational research is not a technique that has received very wide use among poltiical scientists. Since the inception of contemporary empirical work in political science, the major data-gathering technique has been survey research. This is understandable, since surveys allow the researcher to collect large amounts of data from numerous respondents in a relatively short period of time. Also, it is a reasonable way of studying questions that involve people's attitudes and information about political phenomena. One weakness of survey research, however, has been its inability to generate reliable information on actual political behavior. At best, the researcher can get data on reported behavior—either the behavior of the informant or the behavior of persons known to the informant. Observational research, conversely, is well suited to the study of actual political behavior; the researcher (or his assistants) know what activities the subject actually was involved in because they saw the subject involved in them. What is more difficult to ascertain, of course, is the meaning that subjects attribute to their behavior, a difficulty also present in survey and other empirical research.

One discipline in which there has been extensive use of observatronal research is ethology. Ethologists observe the social behavior of animals, and generally they analyze their data within an evolutionary framework. Most relevant to political science is the body of ethological research concerned with the internal governance of animal social groups. Ethologists have used nonparticipant observation to study such phenomena as dominance, attention structure, and leadership within animal groups. The same method can be used to study social influence phenomena, such as power and authority, within human groups. This chapter will explore that use in a preliminary way, with special reference to the potential relevance of such research to the study of political socialization.

Observational Research and Political Socialization

Ethological research to date has been strongly oriented toward the observation of the actual behavior of animals in natural or quasinatural settings (Altmann, 1974; Hutt and Hutt, 1970). Emphasis has been placed on the observation of groups of animals rather than on lone individuals, and this has generated a considerable amount of information on social behavior, as opposed to pure biological functioning. What is emerging is a picture of animal societies or groups that have much more complex and varied social organizations than previously thought. At times, the parallels with human social behavior have been striking—at least on a superficial level (von Cranach, 1975). These findings, in turn, have led to the development of a number of theories and speculations about the evolution, biological basis, and characteristics of animal social life. Often, these theories and speculations have seemed, at least to some scholars, to be applicable at least in a limited way to human social or political behavior.

Two general interdisciplinary areas of research have emerged from the efforts of some scholars to explore the relationship between animal social behavior and human social behavior. These areas are usually referred to as sociobiology and human ethology. Up to the present, sociobiology has been predominantly theoretical in orientation, with a major emphasis on speculation about social behaviors, such as altruism and aggression (Barash, 1977; Wilson, 1975; Wilson, 1978). There has been no substantial development toward sociobiologically oriented empirical research on human beings. Some human ethologists have evinced interest in sociobiological variables and theories; this may presage an eventual convergence of the two fields. Any movement toward convergence, however, will have to overcome the reservations that have been expressed by many human ethologists about various assertions of the sociobiologists. Sooner or later, though, some empirical thrust would seem to be inevitable if sociobiology is to continue to develop as a field and eventually to achieve general acceptance in the scientific community.

Closely allied to developments in sociobiology but at the moment still discrete has been the emergence of a group of individuals from various disciplines who have begun to do empirical ethological research on human subjects (Beck, 1978; Chance and Larsen, 1976; Edelman and Omark, 1973; Feldbaum and Christenson, 1977; Masters, 1978; Omark, Fiedler, and Marvin, 1976; Omark, Strayer, and Freedman, in press; Savin-Williams, 1976; Savin-Williams, in press; Strayer, 1978; Strayer and Strayer, 1976; Strayer and others, 1977; Vaughn and Waters, 1977; Zivin, 1977a, 1977b, 1978). Strongly oriented toward fieldwork, most of these human ethologists have drawn their inspiration from both the methods and the theories of an ethological tradition that is dominated by observational research in natural or quasinatural settings.

Current ethological research that is of most obvious relevance to political behavior concerns dominance and attention structure (Beck, 1978; Barner-Barry, 1978, 1979a; Masters, 1978). Much recent work on these two subjects has been based on data gathered from the observation of young children in preschool programs or play groups. The focus has been on the analysis of the dominance pattern or attention structure of the group as a whole. Either tacitly or explicitly, the primary unit of analysis is the interaction between group members rather than the discrete characteristics of group members.

Observational research by human ethologists on such phenomena as dominance and attention structure is politically relevant in that it has a direct bearing on early experiences that many humans have with respect to such political phenomena as power and authority (Barner-Barry, 1978, 1979a). Children in relatively stable peer groups do govern themselves to some degree, at least part of the time. In this sense, they resemble groups of other animals, such as primates, which are also self-governing via some system of superordination and subordination. There are, of course, some extremely important differences between stable groups of children and stable groups of nonhuman primates. For example, most nonhuman primate groups are composed of animals of different ages and levels of physical maturity, many or all of whom have familial ties. The stable groups of small children currently favored for human ethological study are composed of individuals of similar ages and levels of physical maturity with few or no familial ties. While participating in such groups, children learn ways of coping with the power and de facto authority structures that emerge.

Information about such power and authority structures should be inherently interesting to political scientists, because such structures represent naturally arising and evolving micropolitical systems. As such, they are particularly well suited for basic research on political phenomena (Barner-Barry, 1979a). Such basic research is not, however, a major preoccupation of contemporary political science. Information on the micropolitical systems of such typical early childhood groups as those found in nursery schools, daycare centers, and kindergartens is more relevant to the concerns of contemporary political science because of its potential for illuminating some of the core issues of political socialization.

Unquestionably, children in such groups do learn to cope with the power and authority that is exercised by others; some even develop considerable skill in exercising power or authority themselves (Barner-Barry, 1979b). The question, then, becomes whether or to what extent these experiences carry over to condition subsequent political behavior in larger, more explicitly political systems. Do they have implications for adult political behavior?

Given the general thrust of contemporary research in child psychology and developmental psychology, it seems reasonable to hypothesize that the early childhood experiential history of a human being will tend, to a degree that varies with individual circumstances, to have an impact on subsequent politically relevant behavior. In fact, it is much more difficult to find evidence for the proposition that the early social habits and coping skills of a human leave no significant marks on subsequent behavior, either in the political sphere or in any other sphere of social life. Equally, more and more researchers are questioning the proposition that what one becomes as an adult is largely fixed during the first five or six years. Although it makes research more difficult, there is comfort in assuming that reality is more complex than either of these two extremes suggests. That, on the average, there are important residues seems a much more credible working assumption—at least until the time when research permits the matter to be assessed on the basis of a substantial body of sound empirical data.

Longitudinal Research and Interactional Phenomena

The last point raises the issue of the feasibility of longitudinal research directed toward elucidation of these matters. On a superficial level, it may seem that the way to attack this research problem would be to locate and study a large group of very young children with various background characteristics and to follow up this initial study with additional studies done periodically as the children are growing up. This is the classic way in which longitudinal studies on human traits have been done.

If there is an outstanding piece of research of this type, it is the longitudinal study of gifted children that was begun by Lewis M. Terman in 1921 (Terman and Oden, 1959). Using a series of tests culminating in the Stanford-Binet IQ test, Terman sifted through a school population of a quarter million elementary school children and located approximately a thousand who were intellectually superior. When certain other subjects were added to this original group, the total number of gifted individuals who became part of the study was 1,528 (857 males and 671 females). Over the next thirty-five years, Dr. Terman periodically collected information about virtually all facets of his subjects' lives. Although Terman died in 1956, his associates have continued the study with the express aim of collecting data on the entire lives of this group of intellectually superior people (Sears, 1979). Why not use a similar research design to study the origins and development of politically relevant social characteristics, such as behavior with regard to power or authority?

In addition to the usual problems, such as funding and subject attrition, associated with research of this nature, this particular type of longitudinal research would present substantive difficulties. As stated earlier, such politically relevant social behaviors as dominance, attention, power, authority, and influence are interactional phenomena, and an interactional phenomenon does not exist in the absence of a social exchange between two or more people. Thus, one cannot be authoritative unless someone else is acquiescent; a person is not socially powerful unless he or she wields power over at least one other person. Interactional phenomena are, therefore, more ephemeral than attributional phenomena, such as the color of a person's hair or, for that matter, a person's political disposition. When a person walks out of a particular situation, he takes most or all of his attributes with him. His interactions are, in essence, left behind. When he walks into most other situations, his attributes stay basically the same, while his interactions may not because he will be interacting with different people under different circumstances. In other words, under most conditions, situational influences on attributional phenomena are relatively weak. Situational influences on interactional phenomena are, however, much more likely to be present in significant strength. Thus, the approach to studying interactional phenomena must be informed by an interactional perspective.

The difficulties of doing such longitudinal research from an interactional perspective are indicated in the following description of what can be referred to as the interactionist model: "A basic element in this model is the focus on the ongoing, multidirectional interaction between an individual and his or her environment, especially the situations in which behavior occurs. Persons and situations are regarded as indispensably linked to one another during the process of interaction. Neither the person factors nor the situation factors per se determine behavior in isolation; it is determined by inseparable person by situation interactions. This view has the consequence that research has to focus simultaneously on person factors, situation factors, and the interaction between these two systems" (Magnusson and Endler, 1977, p. 4). If actual behavior is indeed a function of a continuous process of multidirectional interaction or feedback between the individual and the situations he or she encounters, then the interactional behavior of any given individual will tend to change from situation to situation. Whether behavior changes in predictable ways is currently open to speculation. Even if we assume that it does, however, the current state of knowledge in the behavioral sciences does not permit predictions of very high accuracy. What are the implications for the issue at hand?

Animal ethologists have found longitudinal studies of interactional phenomena relatively easy to do. First, the life spans of their subjects are, on the average, much shorter than the life spans of humans. Second, the social groups of animals that ethologists study tend to be naturally occurring family-based units, which stay relatively stable over the lifetimes of their members. Therefore, animal ethologists have succeeded in doing many longitudinal studies of animals that are roughly comparable to the Terman study. The best-known is probably Jane Goodall's research in Africa. In contrast, human ethologists have not developed any substantial focus on longitudinal work that is even approximately comparable to Terman's work. Their research has been carried out in situations where relatively temporary groups of children or adults are accessible for observation. The groups are usually not family-based and exist for only a short interval in the subjects' lives, and the members then separate and go their individual ways.

In order to study human interactional phenomena, however, one must study the behavior of persons while they are in groups. When a person is removed from a given group situation, what residue, if any, remains of that person's behavior in the group? Do such residues have implications for the political socialization process? If politically relevant residues do exist, how can they be identified and measured? If we want to do longitudinal studies of people's behavior with respect to interactional phenomena, then at the very least we must try to keep those persons in stable face-to-face groups. The socialization perspective, which involves a multiyear, multiperson endeavor, requires the investigator to keep large numbers of people in stable groups; this is plainly not very practical.

Therefore, if one regards such phenomena as power, influence, dominance, and authority as interactional rather than individual attributes, one cannot readily use a long-term panel study approach. Research designs such as Terman's are of limited usefulness. If we accept the interactionist model, then we must accept the proposition that when a person changes the group environment, he or she becomes in some sense a different person. Also, the internal and external environment of a group changes constantly in very complex and subtle ways. Thus, it would seem that studying the interactional behavior of a large number of persons over a period of time that is significant from a socialization perspective is at least impractical, if not impossible. Large-scale multiyear studies must await further refinements in data gathering and a higher-level of analytical sophistication in behavioral research, not to mention research funding of a magnitude that is not likely to be forthcoming in the near future.

There is another approach to longitudinal research that, in some form, has been used by social scientists from various disciplines (Simon, 1978; Smith, 1975). This is the panel study, which covers a much more modest period of time. Longitudinal study is still defined in terms of repeated observations and the tracing of the nature and amount of continuity and change over time. The time period involved, however, is usually measured in days or months rather than in years. Since social interaction is involved, the outer limits of the time period that can be spanned are the beginning and the end of the existence of the group. Although this form of longitudinal study is not as useful in the search for answers about continuity and change over individual lifetimes or even in maturational periods, it can indicate factors that have a high probability of being involved in more long-term trends.

For example, studying structures of interactional behavior in stable, bounded groups of children, it is useful to be able to identify and analyze perturbations that change the behavior patterns of individuals in significant ways.

What happens, for instance, when a highly dominant or authoritative child is removed from the group? What happens if there is a radical change in the group's environment, such as a movement from predominantly indoor activities to predominantly outdoor activities? What if the major adult supervisor is permanently or temporarily replaced by another adult with a different approach to supervision?

The results of such research might then be extrapolated to similar events that have a high probability of occurring during most people's maturational periods or adult lives. This process would permit the formulation of hypotheses that could then be tested by identifying groups or individuals about to undergo such a perturbation and by observing their relevant interactional patterns before and after the perturbation occurred. This is not as methodologically neat an approach to socialization issues as the more ambitious type of longitudinal study discussed earlier. However, given the problems of ambitious longitudinal work, this seems to represent a reasonable compromise. If fairly definite patterns began to emerge from the data accumulated in these natural experiments, it would then be possible to make more firmly based assumptions about the experiential history of individuals and their subsequent political behavior patterns.

There is at least one other possibility. If one is interested in such phenomena as dominance, attention, power, and authority for their relevance to political leadership, an alternative approach to longitudinal research would require the investigator to observe systematically a very large number of stable groups of young children, preferably in the three- to five-year-old range. On the basis of observational records, the most authoritative, powerful, or dominant children could be identified, and these children would form a panel of subjects who could be studied over time. This would permit the gathering of information on their behavior in a variety of naturally occurring groups and settings. It would also make it possible to identify experiences and characteristics that turn out to be associated with subsequent political behavior of various kinds. This information could then be combined with data generated from other sources, such as psychobiographical research, experimentation, and surveys, to give a multidimensional insight into some important questions regarding political leadership. For example, why do some people decide to run for elective office, while others with similar abilities and backgrounds choose to exercise their leadership or organizational skills in civic or professional organizations or not at all? Why do some people with politically relevant skills and backgrounds remain politically apathetic? How are the skills that lead to political success encouraged and developed?

Conclusion

The kind of longitudinal research discussed above is not likely to become a heavily used technique in political research. The major reason for this is that observational research is extremely time-consuming and expensive.

Also, observation is a skill that is not easily or quickly acquired, and, as with other skills, a certain amount of natural aptitude is essential in order for an observer to be trained adequately. Regarding the rapidity with which data can be gathered and the quantity of data that can be generated, observational research will never be able to compete successfully with survey research. For some purposes, however, it is the preferred research technique. If one is interested in gathering data on the actual behavior of persons in natural settings, no good substitute is immediately obvious.

To date, there has been relatively little research by political scientists on socialization to such basic forms of political behavior as power, authority, and dominance. One problem has been the difficulty of establishing a base point in early childhood. Questionnaires administered to preschool children yield data of dubious validity at best. This is especially true when the information to be elicited is about something as abstract as political power or authority. School-age children are somewhat better subjects for such research; however, while the validity problems are somewhat diminished, they still remain. At either stage, concepts of the political system may vary from nonexistent to distorted to reasonably accurate. Overall, the formal political system is of such low salience to children that it is necessary to question how great an impact it could possibly have on their lives.

Fortunately, however, if one is studying such phenomena as power, authority, or dominance, one is not limited to data on the formal political system. All children live in situations where there are multiple structures of power, authority, dominance, and influence with which they mut learn to cope as they go about their daily activities. Acquired habits of muscle and mind accumulate and modify over the years. These, in turn, are the habits of adult behavior that are most likely to be generalized to the formal political system when individuals find themselves in appropriate situations. They may not be behaviors about which the individual is capable of intellectualizing; for most adults, they may be classified as gut reactions. If so, to ask mature people questions about such relational phenomena may yield data that are only marginally superior to those which can be gleaned from children. In addition, there are all the other related problems of survey research, such as the tendency for respondents to give what they perceive as socially acceptable answers.

If behavior is at issue, then a technique that studies behavior directly is most appropriate. Observational research attempts to record and analyze actual behavior. Information on what people have done is superior to information on what people say they have done or what we project they will do based on their expressed attitudes. (A considerable body of social psychological research demonstrates that the relationship between attitude and behavior is extremely complex and problematical.) That there are problems associated with an observational approach to the study of basic forms of political socialization does not mean, however, that any other method is inherently superior. It simply means that there is a need to improve and supplement the observational methods with such techniques as surveys of experiments. Ultimately, the solution will probably involve some form of relatively sophisticated triangulation. This chapter is intended as an initial contribution to an ongoing discussion of methodology that will eventually result in the refinement of approaches to the definition and study of basic forms of political socialization — behavioral and attitudinal — from cradle to grave.

References

Altmann, J. "Observational Study of Behaviour: Sampling Methods." *Behaviour*, 1974, *49*, 227–267.

Barash, D. P. *Sociobiology and Behavior*. New York: Elsevier, 1977.

Barner-Barry, C. "The Biological Correlates of Power and Authority: Dominance and Attention Structure." Paper presented at the annual meeting of the American Political Science Association, New York, August–September, 1978.

Barner-Barry, C. "The Structure of Young Children's Authority Relationships." In D. R. Omark, F. F. Strayer, and D. G. Freedman (Eds.), *Power Relationships: An Ethological Perspective on Human Dominance and Submission*. New York: Garland Press, 1979a.

Barner-Barry, C. "The Utility of Attention Structure Theory and the Problem of Human Diversity." Paper presented at congress of the International Political Science Association, Moscow, August 1979b.

Beck, H. "Withdrawal from Viet Nam: Four Biobehavioral Studies." Paper presented at the annual meeting of the International Society of Political Psychology, New York, September, 1978.

Chance, M. R. A., and Larsen, R. R. (Eds.). *The Social Structure of Attention*. New York: Wiley, 1976.

Edelman, M. S., and Omark, D. R. "Dominance Hierarchies in Young Children." *Social Science Information*, 1973, *12*, 103–110.

Feldbaum, C. L., and Christenson, T. E. "An Observational Technique for the Naturalistic Study of Group Processes and Social Interactions in Preschool Children." Training manual distributed at annual meeting of the Animal Behavior Society, University Park, Pennsylvania, June 1977.

Hutt, S. J., and Hutt, C. *Direct Observation and Measurement of Behavior*. Springfield, Ill.: Thomas, 1970.

Magnusson, D., and Endler, N. S. "Interactional Psychology: Present Status and Future Prospects." In D. Magnusson and N. S. Endler (Eds.), *Personality at the Crossroads: Current Issues in Interactional Psychology*. Hillsdale, N.J.: Erlbaum, 1977.

Masters, R. D. "Attention Structures and Political Campaigns." Paper presented at annual meeting of the American Political Science Association, New York, August–September, 1978.

Omark, D. R., Fiedler, M. L., and Marvin, R. S. "Dominance Hierarchies: Observational Techniques Applied to the Study of Children at Play." *Instructional Science*, 1976, *5*, 403–423.

Omark, D. R., Strayer, F. F., and Freedman, D. G. (Eds.). *Power Relationships: An Ethological Perspective on Human Dominance and Submission*. New York: Garland Press, in press.

Savin-Williams, R. C. "An Ethological Study of Dominance Formation and Maintenance in a Group of Human Adolescents." *Child Development*, 1976, *47*, 972–979.

Savin-Williams, R. C. "Dominance Hierarchies in Groups of Middle to Late Adolescent Males." *Journal of Youth and Adolescence*, in press.

Sears, P. S. "The Terman Genetic Studies of Genius, 1922–1972." In A. H. Passow (Ed.), *The Gifted and the Talented: Their Education and Development*. Chicago: University of Chicago Press, 1979.

60

Simon, J. L. *Basic Research Methods in Social Science.* (2nd ed.) New York: Random House, 1978.

Smith, H. W. *Strategies of Social Research: The Methodological Imagination.* Englewood Cliffs, N.J.: Prentice-Hall, 1975.

Strayer, F. F., and Strayer, J. "An Ethological Analysis of Social Agonism and Dominance Relations among Preschool Children." *Child Development,* 1976, *47,* 980–989.

Strayer, F. F. "Dominance, Leadership, and Control Rules among Young Children." Paper presented at annual meeting of the International Society of Political Psychology, New York, September 1978.

Strayer, F. F., and others. *Ethological Perspectives on Preschool Social Organization.* Montreal: Department of Psychology, University of Quebec at Montreal, 1977.

Terman, L. M., and Oden, M. H. *The Gifted Group at Mid-Life.* Stanford, Calif.: Stanford University Press, 1959.

Vaughn, B. E., and Waters, E. "Social Organization among Preschool Peers: Dominance, Attention, and Sociometric Correlates." Paper presented at annual meeting of the Animal Behavior Society, University Park, Pennsylvania, June 1977.

von Cranach, M. *Methods of Inference from Animal to Human Behavior.* The Hague: Mouton, 1975.

Wilson, E.O. *On Human Nature.* Cambridge, Mass.: Harvard University Press, 1978.

Wilson, E. O. *Sociobiology: The New Synthesis.* Cambridge, Mass.: Belknap Harvard University Press, 1975.

Zivin, G. "Facial Gestures Predict Preschoolers' Encounter Outcomes." *Social Science Information,* 1977, *16,* 715–730.

Zivin, G. "On Becoming Subtle: Age and Social Rank Changes in the Use of a Facial Gesture." *Child Development,* 1977b, *48,* 1314–1321.

Zivin, G. "The Relation of Facial Gestures to Conflict Outcomes." Paper presented at annual meeting of the International Society of Political Psychology, New York, September 1978.

Carol Barner-Barry, professor of government at Lehigh University, turned to ethological methodology and theory as an alternative to verbal self-report methods in the study of socialization in young children. In addition to her own published research, she has trained and supervised observers in youth care centers as research scientist at Lehigh's Center for Social Research.

*Biopolitical research can draw on contemporary biology without entailing
reductionism—for example, the concept of attention structure
can be used to study media images of presidential candidates.*

Linking Ethology and Political Science: Photographs, Political Attention, and Presidential Elections

Roger D. Masters

Despite popular misconceptions, biopolitical research need not entail a reductionist methodology. The social sciences can use biology as a source of concepts and hypotheses without reducing the explanation of human behavior to
genetic determinism. As I shall illustrate, biology in general—and behavioral
biology (ethology) in particular—can identify variables suited to analysis by
the traditional canons of the social scientist.

The tendency to equate biological analysis with genetic determinism
rests on a theoretical fallacy. Popularizations like Ardrey's *African Genesis* or
Wilson's *On Human Nature* have implied that the biology of human social
behavior is, by definition, the study of genetic causes. Although this conception is shared by exponents as well as critics of sociobiology and ethology

Thanks are due to students at Dartmouth College who have collaborated over
the last few years in the analysis of data. Photographs in the *New York Times* for 1960
were coded by Susan Chess, Stephen Cook, Amy Gonroff, Jayne Heimlich, Margaret
Kimball, Carol Mason, John Paffenbarger, Marianne Parshley, Lisa Robinson, Denise
Ross, Jay Shoifet, Betsey Slotnick, Tim Taylor, Kathy Wholey, and Darrell Wong.

(Caplan, 1978), it ignores the fundamental distinction between biological causes and biological functions.

As should be common knowledge, Darwin's theory of evolution is based on a distinction between more or less random variation within a species and natural selection favoring some varieties over others. In Darwinian theory, the precise causes of variation are necessarily different from the selective pressures that result in survival, death, and differential reproduction. One must therefore distinguish between the genetic and physiological mechanisms that produce a behavior and the selective advantages of that behavior. The philosophically inclined may be reminded of Aristotle's distinction between efficient and final causes. Sometimes referred to as proximate and ultimate causation, these two kinds of cause have generally been distinguished as cause and function (Kummer, 1978). Whatever the vocabulary, the concept of selective advantage or function cannot be reduced to proximate causes, such as genes, without thereby distorting Darwinian theory (Masters, 1979a).

Social scientists should pay especially close attention to the difference between proximate causes, such as genetic influences on behavior or individual learning, and selective or functional advantages. Functionalism is, of course, well known in anthropology and sociology; one thinks of Malinowsky, Ratcliffe-Brown, Parsons, or Almond. In most functional approaches, however, the definition of variables is somewhat ad hoc or circular. Even in biology, this difficulty can never be completely avoided, insofar as a function is always identified by the observer. However, once a biological definition of adaptive functions has been found useful in the study of other species, it can suggest hypotheses for empirical research in human behavior without requiring us to assume that genetic causation is being invoked.

Biopolitics, and, indeed, the social sciences more broadly, thus can and should borrow functional categories from the biological sciences (Masters, 1976b; Schubert, 1973; Somit, 1976; Tiger and Fox, 1971; Wiegele, 1979; Willhoite, in press). At the outset, such concepts and hypotheses make possible a description of relationships that are otherwise often ignored or misunderstood. After we have a good descriptive understanding of a phenomenon, and only then, does it make sense to seek proximate causes. This argument has been presented elsewhere with reference to the concept of inclusive fitness developed in sociobiology (Masters, 1979b); here, I will focus on ethology.

Ethology: Dominance, Nonverbal Communication, and Attention Structure

The study of animal behavior, now called ethology, was long divided into two hostile traditions (Masters, 1976a). In Europe, particularly under the influence of Konrad Lorenz and Niko Tinbergen, classical ethologists studied instincts or fixed action patterns; researchers in this tradition sought to explain the functional or adaptive significance of the behavior characteristic of a particular species. Meanwhile, particularly in the United States, comparative psychologists used laboratory experiments to explore the precise causation of animal learning; from this perspective, behaviorists like B. F. Skinner sought ubiquitous laws of behavior that would apply equally to most if not all vertebrates.

Contemporary ethology has overcome this divergence as a new generation of researchers adopts subtle combinations of experimental and observational methods (Schubert, 1979). It is widely accepted today that every species has a more or less characteristic range of behavioral responses; due to the recognition of innate differences in perceptual and behavioral repertoires, simplistic S–R models have generally been abandoned. Simplistic theories of instinct have also been given up, because we now know that animals acquire responses in complex ways dependent on individual development and ecological setting. In the old instinct-versus-learning debates, both sides were partly right; hence neither was adequate as a general theory of behavior.

Studies of primates have been of particular importance in the latest generation of ethological research. Lorenz and Tinbergen worked primarily on birds and fish like the greylag goose and the three-spined stickleback. Behaviorists analyzed pigeons, mice, and rats. Since the 1960s, studies of animals in the wild have been extended to species with complex social behavior, including not only lions and hunting dogs but our closest evolutionary relatives, the primates (Altmann, 1967; De Vore, 1965; Kummer, 1978). Inevitably, ethology has changed, if only because the behavior of monkeys and apes is so highly variable, not only from one species to another but also from one group to another within a single species (Rowell, 1969).

The category called *dominance* is a good illustration of the latest developments in ethology. Traditionally, dominance was treated as the result of aggressive but not violent behavior (Altmann, 1967; Lorenz, 1966). For Lorenz, other animals rarely if ever engaged in violence, at least in the wild, because their aggressive instincts were channeled into ritualized gestures symbolizing dominance and subordination. Threatening gestures thus permitted the dominant or alpha animal to gain privileged access to scarce resources or females, while subordinates avoided violent attacks by postures and behaviors signaling an acceptance of inferior status.

While such events are often seen in free-ranging primates, they do not tell the whole story. Since both chimpanzees and langurs have been observed to kill each other (Goodall, 1979; Sugiyama, 1967), it is now obvious that dominance hierarchies in other species are not always maintained without conflict and violence; aggressive behavior sometimes results in sudden shifts in social organization—described by ethologist Michael Chance as "revolutions." Equally important, however, has been observation of social leadership that is not directly related to aggressive behavior at all. In short, relations of dominance and subordination now appear to be much more complex than was thought by those who tried to reduce them to ritualized aggression.

As ethologists explored such complexities, they found that similar behavior could be observed in primates and in human infants or young children (Barner-Barry, 1977; Strayer, 1978, and this volume; Zivin, 1977). In both cases, small face-to-face groups interact in ways that differ from group to group and from one environment to anothe₁. Of particular importance are the nonverbal gestures that serve as social signals. By observing these behaviors very closely, ethologists have found that techniques and concepts of analysis developed to study primates also apply to what is now called human ethology (von Cranach and others, 1979).

Since the publication of Chance's "Attention Structure as the Basis of Primate Rank Orders" (1967), the concept of attention structure has been increasingly useful in such studies. In many primate and human groups, the leader is the focus of attention. Thus, dominance is often, although not always, correlated with the receiving of attention and subordination or following with the giving of attention. Chance and those who have built on his work discovered that such patterns of attention are not always based on aggressive gestures or threats. Rather, a different category of behaviors (called hedonic because it is characterized by display rather than threat) is often associated with leadership (Chance, 1976b, 1977; Chance and Larsen, 1976).

Such research reveals that, in groups of primates and of human children, dominance is highly correlated with attention structure. Moreover, depending on the species and the situation, either agonic behavior (threats and submission), hedonic display, or both may be associated with leadership roles. For example, studies in France (Montagner, 1978; Montagner and others, 1975), Canada (Strayer, 1978 and this volume), and the United States (Barner-Barry, 1977) show that the dominant child in a preschool play group, like the dominant male in a group of monkeys, is not typically the most aggressive individual. Rather, the leader is usually the focus of attention, often engaging in hedonic or reassuring behavior. In many groups, it is the number two male, not the leader, who is the most frequently aggressive.

These findings obviously destroy the popular conception of human ethology as the study of a "naked ape" whose behavior is supposedly determined by "innately aggressive drives." On the contrary, ethology illuminates the differences between humans and other species. Even very young children have an exceptional ability to vary the mixture of agonistic and hedonic modes of attention from one setting to another (Omark and Edelman, 1975). Although most primates have highly developed capacities for monitoring social attention (Kummer, 1978), our species is thus distinguished by the relative fluidity of its modes of social interaction. Humans can shift their focus of attention rapidly, both from environmental objects to the group and from agonic to hedonic modes (Chance, 1977). Indeed, the evolution of the large human brain is probably explained by the selective advantages of this ability to monitor complex social behavior, rather than by the presumed but usually unproved benefits of cognitive ability (Humphrey, 1976).

Attention Structure and Learned Behavior

Ethological research now in progress on both sides of the Atlantic touches on a wide variety of topics, both in primate and human behavior, related to attention structure. On the one hand, observation of social behavior reveals the fruitfulness of reconceptualizing group interaction in these terms (Chase, 1980). On the other hand, Chance's original conception can be related to such fields as cognition, information processing, socialization, and political behavior (Barner-Barry, 1979; Beck, 1976; Chance, 1977; De Reuck, 1980; Masters, 1978; Willhoite, 1980).

Before turning to methodological issues in the restricted sense, it should be explained why the concepts involved can apply to learned or culturally variable traits. Chance's work emphasizes the difference between attention structures based on agonic (threat) versus hedonic (display) behavior. But he also stresses the difference between centric groups, in which individuals all look at one member, and acentric structures, in which individuals attend to distinct microenvironments. None of these concepts necessarily implies genetic or physiological causes.

Good teachers, like trained actors, learn techniques of holding the attention of an audience. So, of course, do politicians (Larsen, 1976). Moreover, cultural norms obviously influence appropriate behaviors. In some societies, a leader periodically gives a four- or five-hour speech outlining fundamental policy operations; in the U.S.S.R. or in Cuba, such lengthy orations are an essential means by which leaders focus the party's attention on basic priorities. In contrast, an American politician who gave a speech of similar length would probably lose his audience irretrievably.

Leaders must also learn the extent to which playful or hedonic behavior is appropriate in different social contexts and the degree to which overt threats are an appropriate social response. An American politician who insisted on the deference accorded to a dictator like Hitler—for example, flying into a rage whenever contradicted and punishing offending subordinates with imprisonment or death—would very likely not be elected. It is from culture, therefore, that one learns the contexts in which it is appropriate to kiss babies and the contexts in which one threatens a subordinate if he fails to do his job.

Since different patterns of monitoring social attention have been found to provide selective or functional advantages in primates, might not the same hold true for humans? If leadership is correlated with serving as the focus of attention in groups of monkeys, apes, and children, does the same relation hold in political campaigns? To test this hypothesis, one must first be able to describe human societies in terms comparable to those already developed in ethological research. At this stage, therefore, it is far more important to observe with a fresh eye and to correlate social attention with leadership than to seek a rigorous causal explanation.

The following discussion of methodology, therefore, leaves the question of causation completely open. Indeed, it assumes that the most frequently discussed phenomena of political attention are matters of cultural variation and individual learning. To be sure, in some cases these factors are not entirely conscious; elsewhere, I have spoken of the importance of gut reactions or vague feelings toward political figures (Masters, 1976a). However, even such emotional responses are doubtless largely conditioned, at least in the precise form of their expression.

Looking at New Aspects of Familiar Data

American presidential elections, like those in many Western democracies, involve competition between candidates. In this process, rivals seek coverage in the press, radio, and television as a means of becoming visible. Name recognition is often considered more important than even a favorable image. Through the successive stages of primary elections, national nominating conventions, and the final campaign, candidates seek to demonstrate their dominance by gaining and keeping public attention. Chance's theory of attention structure can thus provide an interesting way of approaching presidential campaigns (Masters, 1978; Willhoite, 1980).

Under modern conditions, this rivalry for public attention is largely mediated by the press, radio, and television; as a result, the media are often described as the fourth branch of government. An ethological approach therefore suggests that photographs and other visual images of candidates could have more importance than is generally recognized. Precisely because these components of the campaign process are nonverbal, they should be subject to analysis from an ethological perspective. This process should make it obvious that the contemporary importance of media coverage is a cultural phenomenon. It would be absurd to assume that *Time* magazine was caused by the genes of its editors, owners, or readers!

Nonverbal Gestures in Politics: The Candidate as Actor. The first step in analyzing the nonverbal or visual components of the rivalry for political attention is conceptual. That is, one must begin by reconsidering the behavior to be observed. Traditionally, the social sciences have focused on verbal behavior: speeches, party platforms, or public opinion surveys. Often, indexes and variables related to these cultural phenomena—socioeconomic status, age, sex, and the like—are also studied. An ethological approach begins from the injunction that we look, in the literal sense of using our eyes, at politicians and candidates.

This focus immediately indicates the importance of what Goffman (1959) has called the presentation of self: different public figures come across in quite different ways. Elsewhere, for example, I have cited a noted conservative's negative response to a personal meeting with Nixon, long before the latter resigned; in this case, something about the bearing of the man alienated a supporter (Masters, 1976a, p. 216). For most of us, of course, face-to-face meetings with candidates are the exception; our image or sense of the looks of political rivals depends on the media.

The phenomenon to be studied, therefore, is the way in which the political candidate appears to the electorate in a modern election. But this phenomenon is itself the consequence of a series of interrelated processes that it is well to distinguish. First is the candidate, a human individual who gets up in

the morning, shakes hands in front of a factory, sweats under the lights of a television studio, talks with staff aides while drinking coffee, falls asleep exhausted
on an airplane, and so forth. Representatives of the media—reporters focusing
on verbal substance, photographers transmitting visual images—select a sample of this universe of behaviors; editors make a further selection from that
sample of what is to be produced as news. Finally, citizens receive the media
output, paying greater attention to some items, skimming or ignoring others,
and occasionally retaining impressions that will underlie long-term attitudes
and voting responses.

In the process by which the universe of candidate behavior is sampled
and transmitted to the public by the media, there are several distinct levels of
selective attention. The candidates themselves, of course, behave differently,
depending on the anticipated audience. Both the photographer and the editor
are likely to select those photographs that seem to represent the actual behaviors of candidates accurately and that also confirm popular expectations.
Hence, photographers and editors are simultaneously observers influenced by
the candidates' nonverbal and verbal behaviors and producers of media
images that influence the public's response. In preparing images for media display, journalists must combine their desire to present a faithful picture of what
really happens with techniques known to maintain public attention in order to
preserve newspaper circulation or television ratings.

In this process, the candidate is often in the role of the television or
movie actor. Increasingly, of course, this is quite literally the case in the stag-
ing of media events. Short of artificial activities created uniquely to be filmed
for the six o'clock news, candidates are advised on how to walk, stand, and
move. The disastrous public image created by Senator Muskie's sobbing epi-
sode in 1972 demonstrated—if such a demonstration were needed—that a bad
scene can destroy a candidate (Masters, 1978).

There are doubtless personality differences between candidates in their
expression of nonverbal cues, just as there are other differences of character
and style. Jimmy Carter's smile, which exhibits upper and lower teeth, differs
sharply from George McGovern's typical smile, in which the lower teeth are
rarely visible; like Teddy Roosevelt's, the Carter image (at least in the 1976
campaign) was inseparable from this high-intensity smiling gesture. Whether
or how such responses are learned, however, is irrelevant to the method suggested here; while such personal differences can be studied, they are accessible
only to direct observation that bypasses the media. Here, the focus will be on
media images, on the assumption that many politicians are fully conscious of
the importance of their visible images and that they modulate their nonverbal
gestures much as any actor learns to do.

Photographs in Print Media: Taking Politicians' Images Seriously.
In the analysis of nonverbal cue transmission from politicians to the citizen body, it is therefore not promising to try to discover the actual universe
of the candidate's behavior. Rather, it is reasonable to begin from the most
accessible point in the process that links political actors and mass audience.

Visual images in the media obviously serve this purpose well. For the sake of simplicity, photographs in print media have been chosen in preference to television or movie sequences.

Two reasons for stressing photographs in print media may be mentioned. First, they are far simpler to analyze. There is little point in trying to run before one knows how to walk. Hence, before attempting to analyze the complex and rapidly changing sequence of video or film images, still photographs should be studied. Second, there is every reason to believe that the pictures in the print media are an effective means of communicating nonverbal images. Precisely because these pictures do not move, they focus attention in a very different way than the evanescent sequences on television. Even the bystander on the subway or the casual visitor to the doctor's office may be influenced by a photograph on the cover of *Newsweek* or the front page of the *New York Times*.

Since politicians pay considerable attention to their "images," these photographs can legitimately be viewed as an important part of the political process. In so saying, one expects that individual voters need not always respond to a given picture in exactly the same way. On the contrary, just as partisans and opponents respond differently to the same speech, judgments or gut reactions to photographs presumably vary from one person to another. However, before we can understand how these responses to visual stimuli differ, it is necessary to understand the stimuli themselves.

The method described below is intended to permit careful analysis of the photographic images of candidates in electoral campaigns. In so doing, it can be assumed—as a provisional hypothesis and nothing more—that particular kinds of images will elicit predictable responses. This is not a totally unfounded assumption; as a decade of research by Paul Ekman has demonstrated, some nonverbal gestures are decoded in a highly reliable way across virtually all known cultures (Ekman, 1979; Ekman, Friesen, and Ellsworth, 1972). Hence, it is not implausible to assume that some nonverbal features of candidate photographs will stimulate similar emotional reactions, regardless of partisan preference. It should also be emphasized that this is merely a hypothesis and one that should be subjected to empirical test in the future.

Equal Time and Unequal Exposure: Counting as a Methodological Innovation. The first aspect of photographic images that is capable of influencing the electorate is overall frequency. How often have potential voters seen the candidate's face? It is doubtless true that supporters pay more attention to the picture of "their" candidate than to pictures of rivals, but Chance's concept of attention structure suggests the working hypothesis that there should be a correlation between the aggregate frequency of pictures of each candidate and the candidate's dominance as measured by public opinion polls, voter support, and election results.

This hypothesis does not, of course, necessarily imply that having one's picture in the newspaper causes electoral success. That may, of course, happen. But it is also to be expected that dominant individuals will be represented photographically because they are dominant. In other words, the correlation between dominance (or popularity) and attention (or frequency of photographic coverage) reflects a complex interaction in which feedback is extremely important. If a candidate is already popular or holds an incumbent role, he ﹐﹐nds it easier to get coverage than if he is an unknown. Once a candidate begins to be the focus of attention, it is usually easier to increase popularity. Once a candidate loses popularity, the ability to get one's picture in the press also suffers (Masters, 1978).

The hypothesized correlation between popularity or success and photographic coverage in the press is thus easily tested without presuming that it ought to rest on a simplistic causal arrow from attention to popularity or from popularity to attention. The methodology of the test is simple: all one needs to do is count the number of pictures of the candidates that appear during an election campaign. Although raw picture counts can be made more sophisticated by weighting for size of placement, preliminary research suggests that little is gained by the added complexity. Thus, one of the most interesting methodological innovations may consist in simply counting pictures of politicians.

In effect, media images are an easily measured political phenomenon that has too long been ignored. Even if candidates have equal time on television, they may have unequal exposure in the press. Figure 1 indicates, for the presidential campaign of 1972, the percentage of pictures of Nixon and McGovern by stage of the campaign in four national print media (*New York Times, Newsweek, Time, U.S. News and World Report*). The data show that four different publications follow a similar pattern, confirming the ethological hypothesis that attention is correlated with leadership. Note, however, that the four publications diverge in relative emphasis, which suggests that editorial policy also influences the selection of images.

Of particular interest is a pattern common to all four publications. During the critical phase of the campaign, more than 50 percent of the photographs of the two contenders showed Nixon. McGovern dominated the media at the moment of the Democratic convention in July but soon lost that position and was unable to regain it. Even the *New York Times,* which endorsed McGovern, printed more pictures of Nixon from October to November, 1972. Hence, the tendency to be a focus of attention does, indeed, seem to be correlated with political success and leadership (Masters, 1978).

The one-sided pattern in 1972 was in no way inevitable, as illustrated by counts of photographs in the 1960 election. Figure 2a analyzes the number of pictures of Nixon and Kennedy in the *New York Times* by month, and Figure 2b reports the percentages of those favoring Nixon and Kennedy in Gallup polls. There is a striking correlation between political attention, as measured by photographic coverage, and popular support for candidates. In 1960, however, the lead changed hands several times. In particular, Nixon's late burst of momentum—symbolized by the greater frequency of his pictures in the last week of the campaign—coincided with a sharp increase in his support as he nearly overcame Kennedy's lead.

Figure 1. Photos of McGovern as Percent of All Photos of McGovern and Nixon in Four National Publications (by Phases of Campaign, 1972)

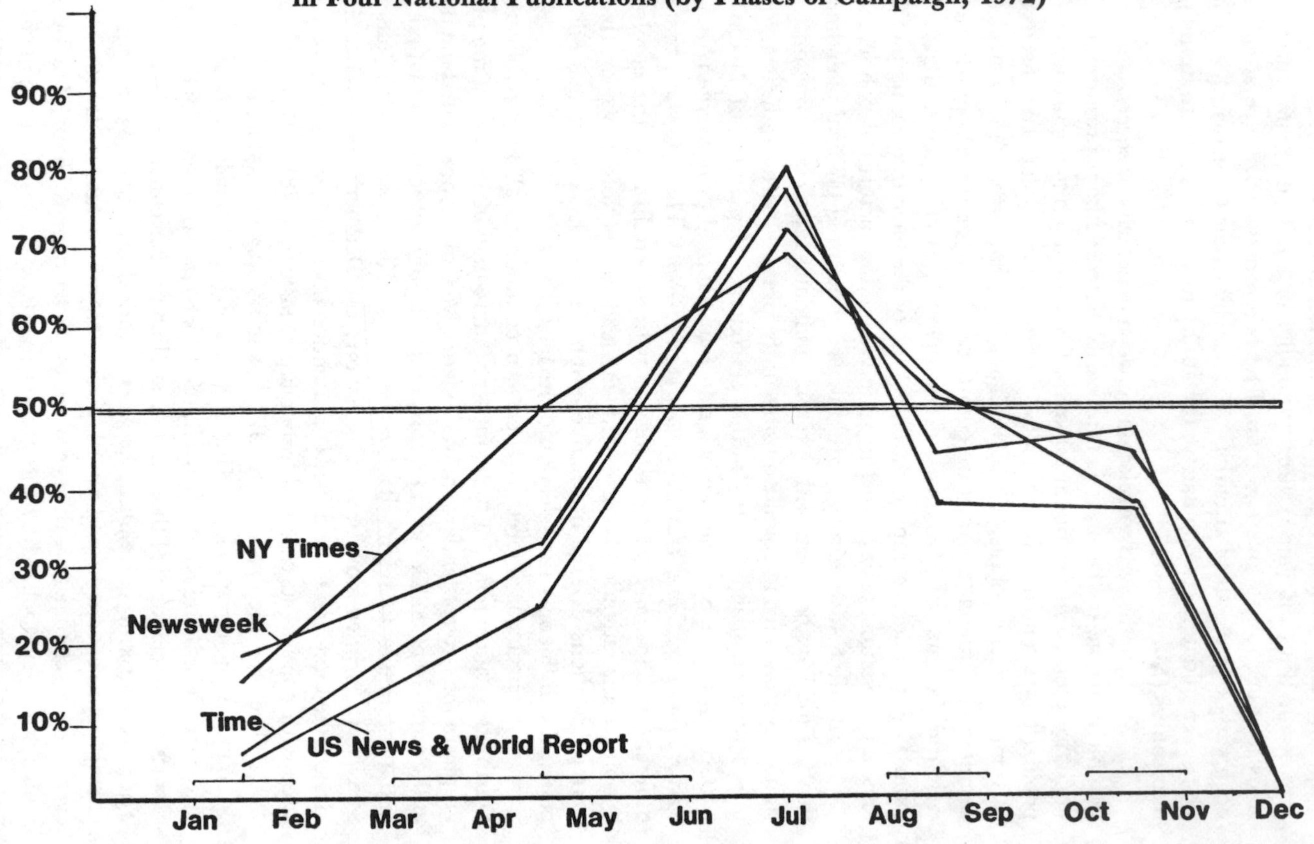

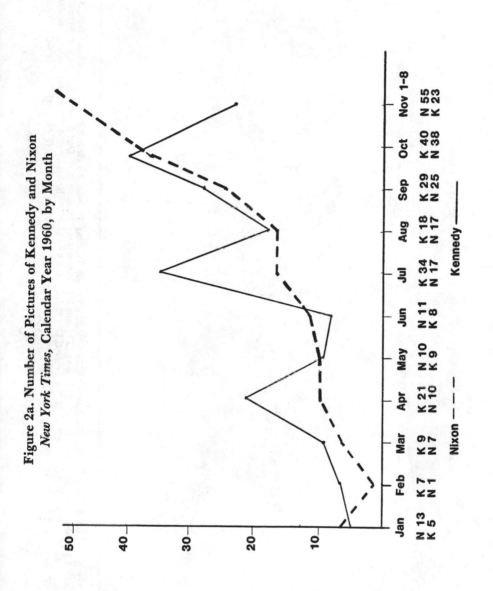

Figure 2a. Number of Pictures of Kennedy and Nixon
New York Times, Calendar Year 1960, by Month

	Jan	Feb	Mar	Apr	May	Jun	Jul	Aug	Sep	Oct	Nov 1-8
	N 13	K 7	K 9	K 21	N 10	N 11	K 34	K 18	K 29	K 40	N 55
	K 5	N 1	N 7	N 10	K 9	K 8	N 17	N 17	N 25	N 38	K 23

Nixon — — —   Kennedy ———

Figure 2b. Popularity Polls, 1960 Presidential Election
Nixon and Kennedy

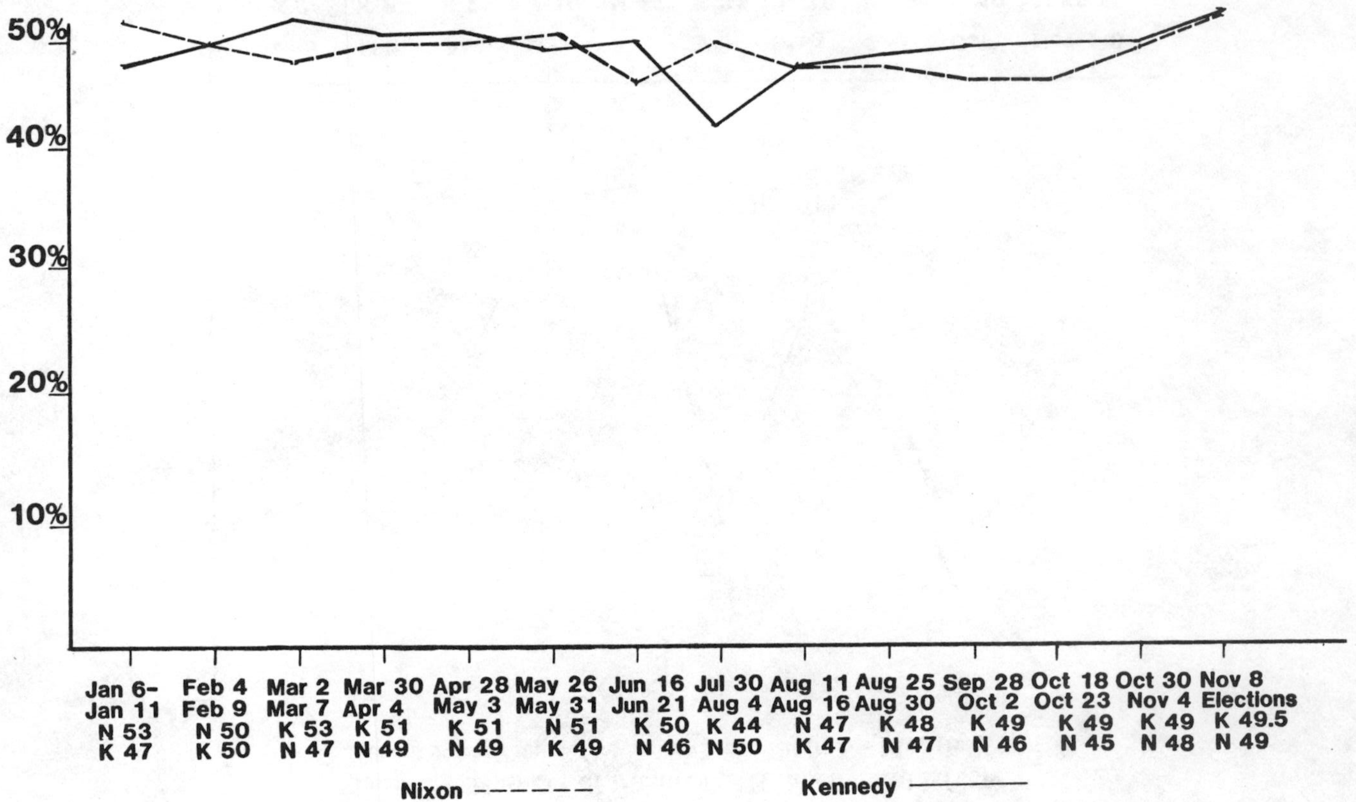
50%
40%
30%
20%
10%
Jan 6– Jan 11 N 53 K 47
Feb 4 Feb 9 N 50 K 50
Mar 2 Mar 7 K 53 N 47
Mar 30 Apr 4 K 51 N 49
Apr 28 May 3 K 51 N 49
May 26 May 31 N 51 K 49
Jun 16 Jun 21 K 50 N 46
Jul 30 Aug 4 K 44 N 50
Aug 11 Aug 16 N 47 K 47
Aug 25 Aug 30 K 48 N 47
Sep 28 Oct 2 K 49 N 46
Oct 18 Oct 23 K 49 N 45
Oct 30 Nov 4 K 49 N 48
Nov 8 Elections K 49.5 N 49
Nixon ------
Kennedy ——

In both 1960 and 1972, therefore, the correlation between political attention and electoral success predicted by ethology is confirmed by the evidence of journalistic photographs. Ultimately, closer analysis may be able to determine the sequential relationship of changes in media attention and public opinion as measured by polls. However, even without studying the causal process by which media coverage influences and is influenced by candidate popularity, the method suggested here can contribute to our understanding of the political process.

Componential Analysis of Photographs

Gross measures of total photographic coverage, while interesting, do not reveal substantive differences in the images of candidates conveyed by pictures. To get at such differences without having to rely entirely on subjective judgment, photographs can be analyzed using methods inspired by ethological research on the facial gestures of primates (van Hoof, 1969) and humans (Ekman, 1979; Ekman, Friesen, and Ellsworth, 1972). Animals and humans seem to respond to overall gestalts (Omark, in press) like the smile face so dear to physicians and dentists. Each of these overall patterns is comprised of elements or components that can be combined in numerous blends, in part because the musculature that can produce each component is distinct. While humans seem to have an overall sense of a smiling or happy face, a sad face, and so forth, the components of these gestures can be and often are recombined in different ways (Ekman, 1979).

As a result, it is risky to attempt overall estimates of whether an individual picture is flattering or insulting. We can all remember animated discussions in which family members attempt to agree on which photograph is the "best" picture of So-and-So. Given the range of subjective variability, photographs should be coded in terms of specific components of the candidates' nonverbal gestures and postures. Only in this way can we research the difference between the photographs as stimulus and actual responses to it.

The Procedure of Coding. The detailed protocol for coding photographs includes two fields or classes of attributes. Field A (so called merely for convenience) refers to three aspects of the photograph as a whole; each of these aspects (or subfields) — the name of the candidate, the date of the publication, and the page on which the photo appears — is coded for each picture. This permits computer search and classification of photographs on any one or more of these attributes.

The second field or class of attributes (Field B) includes 114 descriptors or elements of each picture. Group of descriptors refer to alternative features, only one of which can normally be present in a given image; such groups have been called subfields and are identified by numbers preceding the definition.*

*For reasons of space, a number of illustrative figures and data displays could not be included. Although the text should be clear and self-explanatory, the interested reader is invited to consult the author for operational details of the analysis and coding procedures.

For example, subfield 1, including descriptors 1-3, concerns the position of the eyebrows and the forehead: B1 = eyebrows and forehead raised; B2 = eyebrows and forehead normal; B3 = eyebrows and forehead lowered.

It will at once be asked why the more complete but somewhat more complex coding system developed by Ekman (FAST: Facial Affect Scoring Technique) has not been used here (Ekman and Friesen, 1976). The answer is twofold. First, many of the photographs are either small themselves or they contain images of candidates that are very small. In newsprint, small images have poor resolution. As a result, fine judgments may be difficult for the coder to make; in any case, they are presumably not relevant for the average citizen. Second, many items in Field B are not components of facial gestures, but they still may have significance and hence are worth coding.

In developing a coding system of this sort, moreover, one is not entirely sure of the range of hypotheses to be tested on the data. Hence, it is also desirable to encode some dimensions that might be important on the chance that the original definition of components is not fruitful. This explains the inclusion of such items as Ekman's affect categories. These categories are overall impressions of facial emotion that have been found to be relatively reliable in cross-cultural experimentation. Hence, it seemed interesting to see whether coding for general affect would produce results different from those of componential analysis.

Some additional coded items concern the composition and setting of the photographs. For example, it is well known that a camera angle looking up at the face is subtly more favorable than one looking down (subfield 14). In ethological terms, pictures with the head above camera level (B74) simulate a perception from the position of a subordinate and presumably transmit a cue that the individual photographed is dominant; conversely, pictures with the head below camera level (B76) seemingly indicate that the subject has subordinate status. In a similar manner, pictures showing a profile, half-profile, or the back of the head could well leave a very different impression than head-on photographs (subfield 15).

In addition to such aspects of photographic composition, some elements of the scene itself may be significant. For example, subfield 19 concerns others in the photograph, permitting analysis of the proportion of pictures showing a politician with another nationally known figure in the picture (B100), with another individual (B101), with a crowd (B102), or alone (B103). Media coverage of candidates can differ with regard to these features in interesting and revealing ways, as will be shown below.

One subfield deserves particular explanation, since it might otherwise be difficult to understand. Subfield 18 (Hall's distance) is a scale derived from Edward T. Hall's *The Hidden Dimension*. Hall has shown how changing distance changes one's perceptions in a number of sensory modalities. A photographic image is peculiar because it mimics the retinal image of a specific distance (which, without cropping, might be correlated with the original distance from camera to photographic subject).

It seemed worth considering whether photographs of candidates vary along this dimension. For example, if one candidate is always shown in long shots with a minuscule head, while another is shown in a fully recognizable closeup, we could hypothesize that the latter will receive more attention from the viewer and be perceived as a leader. To make it possible to test such hypotheses, the dimension of distance was coded in terms of the visible facial features, classified along the dimension that Hall calls "60-degree scanning."

Method of Analysis and Illustrative Findings. A computer program has been developed to permit cross-referencing of any combination of the descriptors in either Field A or Field B. Hence, it is possible to go from simple counts of a candidate's number of pictures to searches of complex combinations of gestural components or elements of the setting. Although detailed findings will be reported elsewhere, the method can be illustrated by a preliminary study of politicians' pictures appearing in the *New York Times* during the calendar year 1960. In this data set, 1,685 photographs of 210 politically known individuals were coded and analyzed.

First, several components of the high-intensity response I have called the triumph gesture (Masters, 1978, Figs. 1–3) were identified: eyebrows and forehead raised (B1), eyelids fully open (B4), crow's feet present (B12), teeth wide open (B14), and arms above the head (B42). Since this victory gesture is often exhibited by winning athletes and politicians, it was presumed that its components would be observed more often in pictures of the presidential nominees than in images of other politicians and that the winner in 1960 (Kennedy) would have been shown exhibiting such cues more often than the loser (Nixon).

Analysis was also carried out of the percentage of photographs in the *New York Times* containing four, three, two, one, or none of the high-intensity components. Contrary to the expectation, this apparently dominant cue is more often shown in pictures of other politicians than in photographs of Nixon and Kennedy. Whereas 70 percent of the Nixon pictures and 67 percent of the Kennedy photos include none of these components, only 58 percent of the photos of other politicians could be so classified. At the other extreme, none of the five photographs showing a combination of all four listed cues were pictures of Nixon or Kennedy.

Analysis of a second cluster of gestural components confirms the counterintuitive findings made possible by this method. At the outset, I assumed that a cluster of highly subordinate cues might discriminate between less dominant political figures and leaders. For that purpose, the following components of a submissive gesture (Masters, 1976a, Fig. 15) were chosen for analysis: eyelids slightly closed (B6), eyelids down (B10), teeth not visible (B20), and lips closed (B26). I hypothesized that photos combining all four cues would be relatively submissive and hence likely to be photos of losers.

The relevant test was to examine the frequency of pictures including all four submissive components. As the data indicated, almost all of those shown with these gestures were highly dominant: Eisenhower, Nixon, and Kennedy accounted for fully two-thirds of the pictures in this category. Moreover, half of Kennedy's photographs of this kind occurred after the election, whereas all of Nixon's were prior to election day. In short, it appears that components popularly associated with a highly dominant display are not particularly characteristic of leaders, while submissive or appeasement components are often shown in pictures of highly successful leaders.

The data were thus contrary to intuitive expectations, but they fit reasonably well with the research in primate and human ethology cited above. As in groups of monkeys and preschool children, the leader is less likely to display high-intensity agonic gestures than to display appeasing or facilitating ones. This conclusion was confirmed by the coding for Ekman's categories of facial affect (for example, Figure 3). The pattern or gestalt of a happy face is one of the most universally understood and reliable facial gestures. In 1960, 47 percent of the *New York Times* pictures of Nixon and 44 percent of those of Kennedy were coded as happy, whereas only 39 percent of the entire sample had this appearance. Happiness, not threat or anger, is associated with images of successful leaders.

If these happy or smiling photographs are studied throughout the election campaign, moreover, the shifts in frequency are striking (Figure 3). Through March, the period of his rapid rise in the polls from 47 percent to 54 percent support, Kennedy is often shown smiling. Smiles become predominant for Nixon—and Kennedy's smiles disappear—between late April and early June, the period when Nixon's public opinion ratings rebounded from 46 percent to 51 percent. Nixon's second rebound in the Gallup polls occurred at the beginning of August and again coincided with a peak of happy images, Kennedy's lead in October coincides with an increase in pictures of a happy Kennedy, and Nixon's final burst in the polls during early November coincides with yet another peak of happy faces.

Even though the correlations may not be perfect, the general pattern is striking. This periodicity of pictures showing happy affect is all the more impressive when we consider other facial expressions. Ekman has actually identified a number of relatively standard expressions, including anger, disgust, interest, determination, uncertainty, and so forth. Over the entire campaign, the proportions of these various kinds of facial affect are remarkably similar in pictures of Kennedy and Nixon. Indeed, if one saw only the election year totals, one might assume that there was a standard repertoire of candidate photographs. Yet, as Figure 3 indicates, the ebb and flow of the campaign coincides with marked variations in the frequency of happy images.

Turning from the candidate's facial expression to the setting of the photographs, the method also reveals unsuspected differences in the ways in which politicians are presented. Analysis indicated that Nixon and Kennedy were shown with crowds more than twice as often as other politicians. Presumably, this is a characteristic of a presidential nominee. But if so, does it follow that the frequency of photographs of the candidate in a crowd measures the candidate's successful appeal in a campaign?

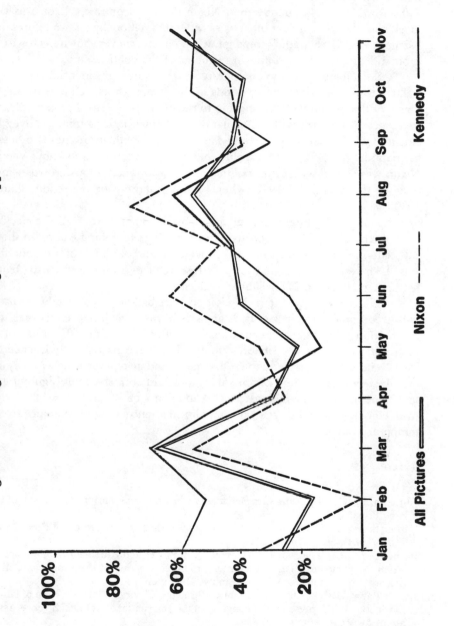

Figure 3. Percentage Photos Showing Ekman's "Happiness"

All Pictures ———   Nixon — — —   Kennedy ———

Furthermore, the percentages of *New York Times* pictures showing politicians in a crowd were plotted by month. Apart from December (when Nixon was shown in only three pictures; in one, he was in a crowd), Kennedy appears with crowds more frequently than Nixon does. In contrast, Nixon was very much more likely to appear in pictures with other nationally known figures than Kennedy was. Early in the campaign, when Nixon is never shown in crowds, he is often depicted with well-known members of the political elite.

Such features as the photographic images in a presidential year might well transmit cues concerning trends in the campaign as well as a nonverbal image of the candidates' relationship to the political system. For example, the frequency of pictures with other nationally known figures might convey the extent to which a leader is perceived as an insider, with strong ties to others in the national political elite. It would therefore be interesting to know whether Nixon was shown with other politicians—or with anonymous crowds—more often in the election of 1972, when he was standing for reelection, than in 1960.

Detailed findings of a comparison between photographs in the elections of 1960 and 1972 will be reported elsewhere. The foregoing examples should suffice to indicate how this method provides a kind of nonverbal content analysis. Like verbal content analysis, the research can limit the study to obvious features, or it can look for subtler patterns.

Needless to say, the methodology described here is perfectly consistent with traditional standards of data analysis in the social sciences. It makes no assumptions about genetic causation, or, rather, it assumes that most of the interesting variation is probably due to cultural and individually learned factors. Neither does it pretend to reduce the candidate to a rat or other lowly animal. Instead, the method permits us to look with a fresh eye at highly important components of the political process and to ask questions that, while derived from ethology, would be relevant in political science without reference to contemporary biology.

References

Altmann, S. A. (Ed.). *Social Communication among Primates.* Chicago: University of Chicago Press, 1967.

Barner-Barry, C. "An Observational Study of Authority in a Preschool Peer Group." *Political Methodology,* 1977, *4*, 415–449.

Barner-Barry, C. "The Biological Correlates of Power and Authority: Dominance and Attention Structure." Paper presented at annual meeting of the American Political Science Association, New York, August 31–September 3, 1978.

Barner-Barry, C. "The Utility of Attention Structure Theory and the Problem of Human Diversity." Paper presented at congress of the International Political Science Association, Moscow, August, 1979.

Beck, H. "Attentional Struggles and Silencing Strategies." In M. R. A. Chance and R. R. Larsen (Eds.), *The Social Structure of Attention.* New York: Wiley, 1976.

Beck, H. "Spiro Agnew the Fool." Paper presented at annual meeting of the American Political Science Association, Washington, D.C., September 3, 1977.

Caplan, A. L. *The Sociobiology Debate.* New York: Harper & Row, 1978.

Chance, M. R. A. "Attention Structure as the Basis of Primate Rank Orders." *Man,* 1967, *2,* 503-578.

Chance, M. R. A. "Social Attention: Society and Mentality." In M. R. A. Chance and R. R. Larsen (Eds.), *The Social Structure of Attention.* New York: Wiley, 1976a.

Chance, M. R. A. "The Organization of Attention in Groups." In M. von Cranach (Ed.), *Methods of Inference from Animal to Human Behavior.* The Hague: Mouton, 1976b.

Chance, M. R. A. "The Social Structure of Attention and the Operation of Intelligence." Paper presented at 139th annual meeting of the British Association for Advance of Science, August-September, 1977.

Chance, M. R. A., Callan, H. M. W., and Pitcairn, T. K. "Attention and Advertence in Human Groups." *Social Science Information,* 1973, *12,* 27-41.

Chance, M. R. A., and Larsen, R. R. (Eds.). *The Social Structure of Attention.* New York: Wiley, 1976.

Chase, I. "Social Process and Hierarchy Formation in Small Groups: A Comparative Perspective." Paper presented at Meeting on Attention Structure in Primate and Human Behavior, Paris, April, 1980.

De Reuck, A. V. S. "Attention Structure in Mediated Conflict Resolution." Paper presented at Meeting on Attention Structure in Primate and Human Behavior, Paris, April, 1980.

De Vore, I. (Ed.). *Primate Behavior.* New York: Holt, Rinehart and Winston, 1965.

Ekman, P. "About Brows." In M. von Cranach and others (Eds.)., *Human Ethology.* Cambridge: Cambridge University Press, 1979.

Ekman, P., and Friesen, W. V. "Measuring Facial Movement." *Environmental Psychology and Nonverbal Behavior,* 1976, *1,* 56-75.

Ekman, P., Friesen, W. V., and Ellsworth, P. *Emotion in the Human Face.* New York: Pergamon, 1972.

Goffman, E. *The Presentation of Self in Everyday Life.* New York: Doubleday Anchor, 1959.

Goodall, J. "Life and Death at Gombe." *National Geographic,* 1979, *155,* 592-621.

Hall, E. T. *The Hidden Dimension.* New York: Doubleday Anchor, 1969.

Humphrey, N. K. "The Function of Intellect." In P. P. G. Bateson and R. A. Hinde (Eds.), *Growing Points in Ethology.* Cambridge: Cambridge University Press, 1976.

Kummer, H. *Primate Societies.* Chicago: Aldine-Atherton, 1978.

Kummer, H. "On the Value of Social Relationships to Nonhuman Primates." In M. von Cranach and others (Eds.), *Human Ethology.* Cambridge: Cambridge University Press, 1979.

Larsen, R. R. "Charisma: A Reinterpretation." In M. R. A. Chance and R. R. Larsen (Eds.), *The Social Structure of Attention.* New York: Wiley, 1976.

Lorenz, K. *On Aggression.* New York: Harcourt Brace Jovanovich, 1966.

Masters, R. D. "The Impact of Ethology on Political Science." In A. Somit (Ed.), *Biology and Politics.* The Hague: Mouton, 1976a.

Masters, R. D. "Exit, Voice, and Loyalty in Animal and Human Behavior." *Social Science Information,* 1976, *15,* 855-878.

Masters, R. D. "Attention Structures and Political Campaigns." Paper presented at meeting of the American Political Science Associaton, New York, August-September, 1978.

Masters, R. D. "Beyond Reductionism." In M. von Cranach and others (Eds.), *Human Ethology.* Cambridge: Cambridge University Press, 1979a.

Masters, R. D. "The Political Implications of Sociobiology." Paper presented to the International Society of Political Psychology, Washington, D.C., May, 1979b.

Montagner, H. "Silent Speech." Horizon BBC2 Program, July 28, 1977.

Montagner, H. *L'enfant et la communication.* Paris: Stock, 1978.

Montagner, H., and others. "Vers une biologie du comportement de l'enfant." *Revue des Questions Scientifiques,* 1975, *146,* 481-529.

Montagner, H., and others. "Behavioural Profiles and Corticosteroid Excretion Rhythms in Young Children, Parts 1–2." In V. Reynolds and N. G. Blurton-Jones (Eds.), *Human Behaviour and Adaptation.* London: Francis and Taylor, 1978.

Omark, D. "The Umwelt and Cognitive Development." In D. Omark, F. F. Strayer, and D. Freedman (Eds.), *Dominance Relations.* New York: Garland Press, 1980.

Omark, D., and Edelman, M. "Comparison of Status Hierarchies in Young Children." *Social Science Information,* 1975, *14,* 87–197.

Rowell, T. "Variability in the Social Organization of Primates." In D. Morris (Ed.), *Primate Ethology.* New York: Doubleday Anchor, 1969.

Schubert, G. "Biopolitical Behavior: The Nature of the Political Animal." *Polity,* 1973, *6,* 240–275.

Schubert, G. "The Use of Ethological Methods in Political Science." Paper presented to the International Society of Political Psychology, Washington, D.C., May, 1979.

Somit, A. (Ed.). *Biology and Politics.* The Hague: Mouton, 1976.

Strayer, F. F. "Social Ecology of the Preschool Peer Group." In W. A. Collins (Ed.), *The Proceedings of the Twelfth Minnesota Symposium on Child Psychology.* New York: Lawrence Erlbaum, 1978.

Sugiyama, Y. "Social Organization of Hanuman Langurs." In S. A. Altmann (Ed.), *Social Communication Among Primates.* Chicago: University of Chicago Press, 1967.

Tiger, L., and Fox, R. *The Imperial Animal.* New York: Holt, Rinehart and Winston, 1971.

van Hoof, J. A. R. M. "The Facial Displays of Catarrhine Monkeys and Apes." In D. Morris (Ed.), *Primate Ethology.* New York: Doubleday Anchor, 1969.

von Cranach, M., and others (Eds.). *Human Ethology.* Cambridge: Cambridge University Press, 1979.

Wiegele, T. C. *Biopolitics: Search for a More Human Political Science.* Boulder, Colo.: Westview, 1979.

Willhoite, F. H. "Some Speculative Comments on 'Attention Structure' in Macro-scale Politics (Especially in the United States." Paper presented at Meeting on Attention Structure in Primate and Human Behavior, Paris, April, 1980.

Willhoite, F. H. "Rank and Reciprocity: Notes Toward a Sociobiological Political Theory." In E. White (Ed.), *Sociobiology and Politics.* Lexington, Mass.: Heath (in press.)

Zivin, G. "Facial Gestures Predict Preschoolers' Encounter Outcomes." *Social Science Information,* 1977, *16,* 715–729.

Roger D. Masters is professor of government and John Sloan Dickey Third Century Professor at Dartmouth College. He comes to the study of biology and politics from the field of political philosophy, having written on Rousseau and traditional political theory. In recent years, he has become interested in relationships between the biological and social sciences, especially the implications of ethology and sociobiology for understanding political behavior.

Biological systems are involved in individuals' responses to socially relevant stimuli.

Individual Differences in Skin Conductance Response to Vicariously Modeled Violence and Pathos

Meredith W. Watts

Even allowing for differences in theoretical emphasis, there seems to be considerable convergence on the general proposition that attitudinal, personality, and physiological processes are jointly implicated in sociopolitical processes (Schwartz and Shapiro, 1973). For example, Bandura suggests that "from a social learning perspective, physiological tension level is determined not only by what one does but by what one thinks" (Bandura, 1973, p. 148). Elsewhere, he argues that internal states, such as the symbolic and imaginal processes, are vital both for storage and retrieval of information and behavioral models and for the arousal and empathetic processes likely to accompany the witnessing of a modeled social behavior. Osgood (Osgood, Suci, and Tannenbaum, 1957; Osgood and McGuigan, 1973) postulates a neo-Hullian process whereby the conditioning of meaning is encoded by the accretion of components or signs of the actual object that elicit portions of the same affect and response as the object itself. He has indicated that paper-and-pencil measures of these components might, in principle, have measurable counterparts in the central and peripheral nervous system (Osgood and McGuigan, 1973).

According to Roessler and Collins, "subject state parallels physiological state" (1970, p. 733). Lipowski states more broadly, "Man's social environment is a source of symbolic stimuli, that is, events and situations which impinge on individuals as information. The latter is processed at the psychological level of organization and endowed with subjective meaning, conscious and unconscious. *The nature of the meaning is determined by the person's individual characteristics, innate and learned,* enduring and current. Depending on the way in which the information is evaluated by the subject, it sets off affective and psychomotor responses as well as physiological changes which provide feedback modifying the cognitive processes whereby information is evaluated. . . . A symbolic process may thus lead to affective and physiological changes both through evaluative, cognitive processes of the cerebral cortex, and by the direct outflow of impulses from the reticular activating system to neural structures controlling all bodily processes" (Lipowski, 1974, p. 407; emphasis added). Clearly, if such statements are to be taken seriously, as I believe they must, theories that seek to integrate these areas should be given serious consideration.

An example worth close examination is the theory elaborated by H. J. Eysenck, who has sought to relate attitudes and ideology to the personality and biological heritage of the individual (Eaves and Eysenck, 1974, 1975). While some of his work lends itself to topically controversial implications (concerning, for example, sex differences or criminality), it is possible to bracket out the ideological detritus from some propositions whose truth value must be evaluated in order to find a proper balance in the interaction of the environment with ontogeny.

This need was indicated by our previous work, which, although it was experimentally inelegant, did stumble over the complexities of personality and attitude in its analysis. First, gross measures of television exposure among children were found to be uncorrelated with skin conductance response (SCR) during viewing of violent or aggressive sociopolitical models of behavior, yet attitude measures were correlated. Moreover, heart rate response showed a predictable pattern of directional fractionation according to the direction of the subject's affect (Watts and Sumi, 1976, 1979; for the theoretical basis, see Lacey and Lacey, 1974). Subsequent research indicated a relationship between ego strength and physiological response variability, which lent credence to Eysenck's proposition that certain personal characteristics are related to responsiveness to the environment (Roessler, 1973; Roessler and Collins, 1970). Machiavellianism—denoting aloofness, manipulativeness, and affective disengagement (Christie and Geis, 1970)—has also been posited as a personality trait and found to be correlated with SCR responsiveness (Watts, in press; Watts and Sumi, 1979).

In brief, Eysenck suggests that "the change which stress makes in performance depends to a very large extent on the prior degree of the subject, that is, his personality" (Eysenck, 1975, p. 446) and that introverts and extraverts differ both in the habitual level of arousal and in their reaction to stimulus input, "introverts tending to a high level of arousal and an amplification of stimulation and extraverts to a low level of arousal and a damping down of stimulation" (Eysenck, 1975, p. 447). J. A. Gray (1972) has proposed that introverts are more conditionable to fear and punishment, and Eysenck finds evidence in the lower pain and auditory thresholds of introverts that they have "weaker" nervous systems (Eysenck, 1972, 1980). The "stronger" nervous systems of extraverts are less arousable and have higher thresholds of intensity before recognition and response. This is generally consistent with our findings (Watts, in press; Watts and Sumi, 1979) that Machiavellianism is associated with decreased SCR response during viewing of videotape models of violent or aggressive behavior.

The current research was originally conceived of as an attitude change experiment, with physiological arousal (as measured by SCR) as the intervening variable. For reasons outlined above, Eysenck's extraversion scale was included, as were three attitudinal measures: Violence Ideology (as used in Watts, in press, and in Watts and Sumi, 1979), Castration of Rapists, and Gun Control. The last two scales used a modified semantic differential technique.

Procedurally, subjects were exposed to audiovisual stimuli on video-tape that were replete with violence and pathos, including both the crimes and the pathetic aftermath for survivors. Much of the material from the NBC special "Violence in America" was focused on sex- or handgun-related violence. Thus, the attitude scales on castration of rapists and gun control were considered to be of critical relevance.

As it turned out, the thirty-three male subjects showed significant attitude change from the pretest to the posttest as a result of exposure to the pathos surrounding violent assaults. The male sample increased from a mean opposition to gun control of 88.3 before the experience to 94.9 after the experience (paired observations t-test significant at .02), indicating attitudinal change as a result of the experimental treatment.

However, the exposure to experimental stimuli was the only variable that was physically manipulated; individual differences were measured for analysis by correlational techniques. Since, "from an experimentalist's point of view, personality cannot be manipulated readily . . . " (Roessler, 1973, p. 315), correlational analysis within subjects was superimposed on the simple pretest-posttest design.

Data Analysis

It would be reasonable, judging from our early studies (Watts and Sumi, 1976, 1979) to hypothesize that such personality characteristics as Machiavellianism and ego strength significantly modify or channel response to important aspects of the sociopolitical environment; so Eysenck's theory would tell us. Therefore, we hypothesized that personality characteristics would be of equal or greater significance in determining the degree of arousal or activation when the individual was presented with models of actual or simulated interpersonal behavior.

Specific stimulus materials were taken primarily from an NBC television special, "Violence in America," and featured a pastiche of violent sports, pictures of child-abuse victims and interviews with remorseful parents (one of whom had actually killed her child), the aftermath of violent gun-related crimes (together with interviews with the victims' relatives), and a dramatic depiction of a shotgun killing from "The Autobiography of Miss Jane Pitman." The last scene had been used in previous studies by this author, and it was known to have consistent arousal value because of the violence of the encounter, its sociopolitical significance, and the high quality of the dramatic presentation. Volunteer subjects were pretested on the paper-and-pencil measures several weeks prior to the laboratory experience. Instrumentation and procedures were identical to those described in Watts and Sumi (1976, 1979). For this study, skin conductance response was measured with finger-cup electrodes and monitored with a Gilson five-channel polygraph. Since our interest in this analysis was the effects of personality on arousability, SCR was the prime physiological parameter analyzed.

It is difficult to maintain a clear distinction between personality and attitude measures. Extraversion, which was measured with the original Eysenck scale, is probably a personality measure; likewise, Violence Ideology is probably an attitude measure, owing to its focus on approval of violent or aggressive interpersonal behavior as a solution to sociopolitical problems (see Watts and Sumi, 1979, for a description of this scale). Castration of Rapists was believed to be an attitude scale, but this domain is included in Eysenck's Toughmindedness Scale (Eysenck, 1954), and thus it shares in both topical attitudinal and personality characteristics. Table 1 shows the zero-order correlations among the three scales.

Castration of Rapists was unrelated to Violence Ideology in general, but it did correlate positively with Extraversion. Extraversion itself also correlated significantly with Violence Ideology. (This is consistent, I believe, with Eysenck's theory.) Once this much had been established, the critical question concerned the association of each measure with the individual's arousal as measured by skin conductance response during viewing of the stimulus material. In Table 2, attitudes toward Gun Control, Castration of Rapists, and Violence Ideology (acceptance of violence) had nonsignificant associations (although the latter two were in the low to moderate range). The only significant correlation was between Extraversion and SCR activity. The observed correlation was significant at the .005 level, and its magnitude of .43 was increased to only .45 by addition of the other scales to the multiple correlation.

Table 1. Zero-Order Correlations Among Attitude/Personality Measures (n = 32 males)

	Casrap	VI
Castration of Rapists	—	—
Violence Ideology	ns	—
Extraversion	.32 (.04)	− .35 (.02)

Table 2. Zero-Order Correlations with Total SCR
(n = 32 males)

Gun Control (Pro)	− .04	ns
Casrap (Pro)	− .18	ns
Violence Ideology (Pro)	.20	ns
Extraversion	.43	.005
Multiple R =	.45	

Table 3 pursues this question further by partialing out the influence of each variable singly. Only Extraversion maintains a significant association when the influence of others has been removed. Conversely, the influence of Castration of Rapists and Violence Ideology virtually disappears when Extraversion is considered. The second-order partial correlations in Table 4 show conclusively that the associations for Violence Ideology and Castration of Rapists decrease to virtually zero, while Extraversion retains a significant correlation when the other two measures are controlled statistically. (Professor David Gow of Rice University obtained similar results using path analysis on these data; I am indebted to him for his insights into the implications of Eysenck's theory.)

Discussion

Elaborate conclusions are neither warranted nor possible here. But these roughly designed and implemented experiments provide evidence that attitude alone is insufficient in both conceptualization and measurement to account for the cognitive and affective aspects of individual response to models of interpersonal social behavior. Analysis indicated that attitude toward gun control did change from the pretest to posttest, but that change was not found to be related to SCR. Thus, contrary to expectation, change was not obviously mediated by arousal. It seems more likely that arousal and activation were mediated by individual personality characteristics. In brief, the more extraverted the individuals, the more likely they were to show low arousal when exposed to models of violent or pathetic interpersonal behavior and the more likely they were to be accepting of aggressive behavior in the solution of socio-

Table 3. First-Order Partial Correlations with Total SCR Arousal

Controlling for CASRAP	
VI	.20 (p = .14)
EXTRA	− .41 (p = .01)
Controlling for VI	
EXTRA	− .42 (p = .01)
CASRAP	− .16 (p = .16)
Controlling for EXTRA	
CASRAP	− .05 (p = .39)
VI	.05 (p = .40)

Table 4. Second-Order Partial Correlations

EXTRA	− .37 (p = .02)
VI	.05 (p = .39)
CASRAP	− .06 (p = .38)

political situations. Thus, there seems to be evidence for a triad composed of attitude (as measured by verbal self-report), personality (as measured by various scales, including Machiavellianism and Extraversion), and physiological arousal.

While Eysenck's theory of extraversion and biological heritability may not necessarily in itself be the theory of choice, there is support for the proposition that biological systems are implicated in individuals' responses to socially relevant stimuli. Overt gross-body behavior was not analyzed in this experiment, but covert physiological behavior seems clearly to be a meaningful and, to some extent, a predictable concomitant of cognitive and affective processes.

References

Bandura, A. *Aggression: A Social Learning Analysis.* Englewood Cliffs, N.J.: Prentice-Hall, 1973.

Christie, R., and Geis, F. L. *Studies in Machiavellianism.* New York: Academic Press, 1970.

Eaves, L. J., and Eysenck, H. J. "Genetics and the Development of Social Attitudes." *Nature,* 1974, *249,* 288–289.

Eaves, L. J., and Eysenck, H. J. "The Nature of Extraversion: A Genetical Analysis." *Journal of Personality and Social Psychology,* 1975, *32,* 102–112.

Eysenck, H. J. *The Psychology of Politics.* London: Routledge & Kegan Paul, 1954.

Eysenck, H. J. "Conditioning, Introversion-Extraversion, and the Strength of the Nervous System." In V. D. Nebylitsyn and J. A. Gray (Eds.), *Biological Bases of Individual Behavior.* New York: Academic Press, 1972.

Eysenck, H. J. "The Measurement of Emotion: Psychological Parameters and Methods." In L. Levi (Ed.), *Emotions: Their Parameters and Measurement.* New York: Raven Press, 1975.

Eysenck, H. J. "Man as a Biosocial Animal: Comments on the Sociobiology Debate." *Political Psychology,* 1980, *2,* 43–51.

Gray, J. A. "The Psychophysiological Nature of Introversion-Extraversion: A Modification of Eysenck's Theory." In V. D. Nebylitsyn and J. A. Gray (Eds.), *Biological Bases of Individual Behavior.* New York: Academic Press, 1972.

Lacey, J. I., and Lacey, B. C. "On Heart Rate Responses and Behavior: A Reply to Elliott." *Journal of Personality and Social Psychology,* 1974, *30,* 1–18.

Lipowski, Z. J. "Psychosomatic Medicine in a Changing Society: Some Current Trends in Theory and Research." In P. M. Insel and R. H. Moos (Eds.), *Health and Social Environment.* Lexington, Mass.: Lexington Books, 1974.

Osgood, C. E., and McGuigan, F. H. "Psychophysiological Correlates of Meaning: Essences of Tracers?" In F. J. McGuigan and R. A. Schoonover (Eds.), *The Psychophysiology of Thinking.* New York: Academic Press, 1973.

Osgood, C. E., Suci, G. J., and Tannenbaum, P. H. *The Measurement of Meaning.* Urbana: University of Illinois Press, 1957.

Roessler, R. "Personality, Psychophysiology, and Performance." *Psychophysiology,* 1973, *10,* 315–327.

Roessler, R., and Collins, F. "Personality Correlates of Physiological Responses to Motion Pictures." *Psychophysiology,* 1970, *6,* 732–739.

Schwartz, G. E., and Shapiro, D. "Social Psychophysiology." In W. F. Prokasy and D. C. Raskin (Eds.), *Electrodermal Activity in Psychological Research.* New York: Academic Press, 1973.

Watts, M. W. "Psychophysiological Analysis of Personality/Attitude Scales: Some Experimental Results." *Political Methodology,* in press.

Watts, M. W., and Sumi, D. "Desensitization of Children to Violence? Another Look at Television's Effects." *Experimental Study of Politics,* 1976, *31,* 1–24.

Watts, M. W., and Sumi, D. "Studies in the Physiological Component of Aggression-Related Attitudes." *American Journal of Political Science,* 1979, *23,* 528–558.

Meredith W. Watts is professor of political science and assistant chancellor, University of Wisconsin–Milwaukee. He has written on political behavior and attitudes and on biopolitics.

Voice stress analysis has produced mixed but promising research findings in a variety of natural and experimental settings.

Methodological Aspects of Voice Stress Analysis

Leonard Hirsch
Thomas C. Wiegele

Voice stress analysis is an attractive research methodology because it allows remote assessment of physiological indicators of psychological stress that previously had been accessible only through direct sensor attachment to a subject. This capability promises to open up new research possibilities for social scientists, ranging from the analysis of individual behavior to a better understanding of group decisional processes and communications.

The Nature of Voice Stress Analysis

Voice stress analysis was developed on the basis of the fact that a change in muscle microtremors takes place in the physiology of the human speech mechanism at times of psychological stress. These tremors are in the range of eight to twelve hertz, inaudible to the human ear, and thus not openly detectable. Voice stress analyzers filter out the audible sections of speech so that the pattern of the microtremors can be displayed visually on a strip chart for easy inspection.* The analysis of the patterns lies at the heart of voice stress research.

*Numerous brands of electronic instruments can be utilized in voice stress analysis. The literature of voice stress analysis has developed almost exclusively around

Two types of vocal changes can result when an individual finds himself under strong psychological stress. Bell, Ford, and McQuiston (1974) identified one type as a gross vocal change that can be detected audibly by noting variations in rate, volume, voice tremor, change in spacing between syllables, and change in the fundamental pitch or frequency of the voice. Some subjects are capable of exercising conscious control over such variations.

The second type of vocal change is not detectable by the human ear and results when an individual is under psychological stress. This phenomenon is unconscious and it cannot be controlled by the subject. It results from a slight tensing of the vocal cords under conditions of minor stress and produces dampening of selected frequency variations. Bell, Ford, and McQuiston (1974) state, "When graphically portrayed, the difference is readily discernible between unstressed or normal vocalization and vocalization under mild stress . . .these patterns have held true over a wide range of human voices of both sexes, various ages, and under situational conditions."

Using electronic filtering and frequency discrimination techniques, voice stress analysis processes the voice frequencies from audiotapes and displays the inaudible stress-related frequency modulation (FM) patterns on a moving strip chart (or digital display on some equipment). The tremor signal is separated by a double demodulation process.

The underlying basis of voice stress analysis, the microtremor, has not been independently identified in the vocal mechanisms. McGlone (1975) tested for the microtremor by inserting an electrode in three parts of a male's vocal mechanism (the posterior coarytenoid muscle, the cricothyroid muscle, and the orbi cularis oris inferior muscle) and subjecting the impulses resulting from speech to an electromyographic analysis. This analysis showed no eight-to-twelve hertz frequencies during nonstress vocalizations in any of the three muscles. He did, however, find the microtremor when he attached the electrode to the bicep brachii, which suggests that the microtremor is only found in the extremities.

This experiment, and some partially corroborating evidence cited by McGlone and Hollien (1976), leads one to question the precise physiological basis of the technique. There is clear evidence of a frequency dampening phenomenon in stressful situations, but this change could be caused by other physiological mechanisms. Therefore, there remain questions about the underlying basis of voice stress analysis for which adequate answers are not yet available. In actual practice, however, the technique has displayed a number of promising characteristics.

Examples of Voice Stress Measurement. Let us examine how stress is represented on a voice stress analysis trace. The normal wave form of a spo-

the device known as a Psychological Stress Evaluator (PSE-101) marketed by Dektor, CIS, Inc., Springfield, Virginia. So ubiquitous has been the use of this instrument that voice stress analysis is frequently referred to as "PSE research." Strictly speaking, this is an improper designation. The present writers have utilized the PSE-101 in their research.

ken word under nonstress conditions would appear as in Figure 1. Each individual wave form, or trace, represents a single word. Note the irregularity of the cycles and that the general configuration resembles a wave; that is, if a line is drawn connecting the top points of each cycle, an irregular pattern will be evident.

Figure 2 illustrates three examples of the charted responses of a person under strong psychological stress. As stress becomes stronger, the undulating wave patterns assume increasingly squarish configurations. The patterns in Figure 2 show a degree of rectilinearity believed to be proportionate to the amount of stress present. Increasing levels of stress are shown from left to right.

The amplitude (height) of the trace is arbitrarily adjusted to fit the tape width of the instrument and is not a factor in interpretation. The positioning on the strip chart can affect the overall configuration of the trace, however, and some experience with the equipment is necessary to find optimum settings.

Review of the Literature. A growing body of literature employs voice stress analysis as a research tool. The results, for the most part, are inconclusive or positive, depending on the nature of the study. Since the voice stress analyzer is often portrayed as a lie detector, one group of experimental studies tests the validity of this claim. These studies have not found the voice stress analyzer to be consistently reliable. The second group of studies examines stress and the effect of stressors on the individual in both applied and experimental situations. With certain reservations, these studies have generally found voice stress analysis to be a useful tool for certain types of social research.

The Deception Studies. Dektor reports four studies in a brochure on the voice stress analyzer. In the first validity study performed by Dektor, analysts evaluated charts of "To Tell the Truth" contestants. Out of seventy-five contestants, seventy-one were correctly ascertained for truth or falsehood (94.7 percent). Second, Dektor ran a peak-of-tension test with twenty-four subjects; twenty-two were correctly identified. The third report by Dektor describes a Maryland police chief who found 100 percent corroboration between

Figure 1. Three Traces of Unstressed Words

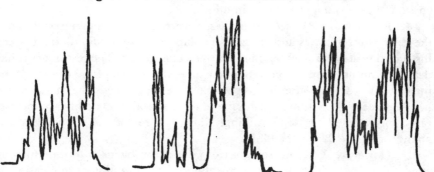

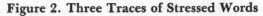

Figure 2. Three Traces of Stressed Words

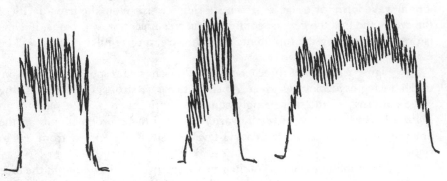

voice stress analysis results and guilt and innocence on 162 elements in questioning of felony suspects (real-world data culled from confession or investigation). Finally, Dektor analysts were able to ascertain correct word groupings for traces of "presumed emotion-producing" or "presumed neutral" words (fifty-two of fifty-three; 98.4 percent accuracy).

Variations on the peak-of-tension and emotive/neutral word tests have been repeated by others but with much less successful results. Lewis and Worth (1975) tested for differences in voice-recorded stress resulting from verbal and written stressors. To do this, subjects were asked to associate to a series of ten written words—five neutral (*string, shoe,* and so forth) and five emotional (*blood, coffin,* and so on)—and to ten spoken words. The five neutral words were spoken normally and as "high-pitched outbursts" (Lewis and Worth, 1975). These observers found that there was no consistent reaction to the written stressors; as often as not, subjects would react more stressfully to a neutral word than to the emotional words. However, they did find substantial difference in the oral experiment, with the loud stressors creating significantly more stressful responses than the normally spoken (identical word) stressors. Thus, they conclude that the manner of presentation, rather than the content, has much to do with the resulting stress level in the subject. To generalize this finding: There are serious questions as to our ability to determine stress level and other psychological states from the content of vocalizations or written communications. What is indicative of stress in some people (for example, discussion of death) may not be so in others.

Horvath (1978) used the peak-of-tension test to compare results from the voice stress analysis and the galvanic skin response (GSR). Subjects were instructed to conceal a chosen number when questioned. The researcher could detect the number only 20 percent of the time, no different from chance. With the galvanic skin response, it was possible to detect correctly the chosen number 70 percent of the time. Furthermore, the intercoder reliabilities of the two physiological methods were strikingly different: r = .38 for the voice stress analysis and r = .92 for the GSR.

However, there are some serious design problems with the Horvath study. The first, which the author mentions, is the question of motivation of the subjects. At the low motivation levels that Horvath analyzed with subjective coding, the range of stress within the traces may not have provided sufficient variance for clear delineation. Second, the peak-of-tension test had only five items, perhaps too short an instrument to control for pre- and postdeception anxiety. Third, Horvath did not analyze the trends of the responses or the placement of the deception within the test. Perhaps those who chose central numbers (2, 3, 4) showed less reliable patterns than those on the ends (1,5). Last, by asking the questions with the numbers ordered in a manner known to the subjects, the placement of the deception was not random but expected and therefore less stressful. Reeves (1976) has shown that the voice stress analyzer is a measure of reaction to a stressor "presented at the moment," and thus foreknowledge should lessen the effect of the stressor at the moment of deception.

While all of these qualifications are valid, they do not explain the different results obtained by voice stress and GSR techniques. It is possible that GSR is a more sensitive indicator at these levels of stress; however, Horvath's study poses more questions than it can answer about the validity of voice stress analysis.

Like Horvath, McGlone (1975) found that identification of deception was not possible at low levels of stress, and he designed an experiment to increase the range of stress. Twenty male subjects read "An Apology for Idlers" by Stevenson while experiencing successively increased levels of shock with randomly determined time intervals between trials. Baseline data words were recorded before the administration of the most severe shock trials. Under these conditions, stress was correctly detected by voice analysis 77 percent of the time.

While deception (or fear of detection) is expected to be a psychological stressor, stess can obviously result from other phenomena. Any issue that might cause anxiety or arousal in a person, such as uncertainty, centrality to desires, or anticipation, should be detectable. Analysts who use this technique to detect deception must be sensitive to the alternative psychological phenomena that can bring forth the stressed reaction.

The Nondeception Studies. There has been no experimental work on the level of stress that is needed for voice stress analysis to become reliable or on whether reliabilities change in a linear fashion as range of stress changes. However, as suggested in the McGlone study, voice stress analysis appears more reliable when the stress level is relatively high.

Brenner, Branscomb, and Schwartz (1979) performed two experiments that support McGlone's study. The first was a deception study in which they tried to increase motivation, thereby raising the stress level, by offering a reward for concealment of a correct response. Subjects were informed of the nature of the study and of voice stress analysis and "were advised that it might be better to produce emotional responses to incorrect items rather than attempt to suppress actual emotional responses to correct ones" (Brenner, Branscomb, and Schwartz, 1979, p. 352). Not surprisingly, Brenner and associates were not able to distinguish the correct items. The study was modeled on earlier research using the GSR that had resulted in positive identification of correct and incorrect items, and thus, it questions the utility of voice analysis at low levels of stress.

In the second experiment, Brenner's group employed a mental arithmetic task where subjects had to perform problems of varying difficulty under a fixed pacing schedule. Prior studies had shown that GSR increased with problem difficulty. Traces of the subjects' voice stress analysis showed linear trends, clearly indicating increasing stress with increasing difficulty. As in the McGlone study, when the range in the stress levels was substantial, voice data conformed to expectation and paralleled GSR results.

Brenner (1974) examined the effects of group size on the intensity of stage fright. In an empty auditorium, speech students from the University of Michigan were asked to recite poetry. Then, they were split into four groups and instructed to recite a very difficult poem before twenty-two, eight, two, and no spectators, respectively, fron an introductory psychology class. The relationship between voice stress analysis scores and self-reports of arousal was high ($r = .32$). Stress increased as an almost perfect function of group size.

Thirty subjects (fifteen male, fifteen female) were asked to read a list of eight one-syllable numbers by Shearer and Wiegele (1977). During the enunciation of words chosen randomly for each subject, a stressor (a burst of high-frequency noise lasting 500 milliseconds in free field 100 db SPL) was presented. Both GSR and voice stress analyses were done by an independent experimenter. The voice analysis traces were correctly chosen for ten of the thirty subjects. This is significantly better than the results of Horvath's comparison of GSR and voice stress.

Smith (no date) conducted two relevant experiments, the first using a six-question format containing two control questions and four sensitive questions. Twenty subjects were asked to agree or disagree with each question. These two control questions dealt with the days of the week. One related to test anxiety and was more stressful than the other. The remaining question served as a stressor. The experimenters asked whether the subject had ever purloined office supplies.

It was found that the mean stress on the first control question was significantly higher than the mean stress on the second control ($p = .05$). The combined mean stress of the relevant sensitive questions was also significantly higher than on the second control question ($p = .05$). The second experiment produced similar results.

Psychiatric researchers have also shown interest in voice stress analysis. Using this technique, Wiggins, McCrainie, and Bailey (1975) analyzed recordings of psychiatric interviews of children. They used three major response categories to the children's answers: content of communication, responses to therapist-posed questions, and miscellaneous responses. Content responses included people, ideas, objects, and actions. Subsequently, analysis of the children's responses to various people in their environment was undertaken. The difference in stress levels between response to direct questions and child-initiated topics was also analyzed. The study found that stress in the voice could be detected and assessed and that voice stress analysis could lead to new ways of exploring the relationship between psychological stress and content of communication in psychiatric patients.

Borgen and Goodman (no date) designed an experiment evaluating the responses of eight paid male prisoner volunteers to a Stroop Test. The Stroop Test produces a conflict situation that is assumed to be stress-inducing. Respondents' skin potential, blood flow, and EKG readings were taken during the test while their verbal responses were being recorded. To test for the effectiveness of antianxiety drugs, subjects received either ten milligrams of diazepam, a tranquilizer, or a placebo two hours prior to examination.

The Stroop Test effectively elicited an arousal state that was noted by significant increases in blood flow, skin potential, and systolic and diastolic blood pressure levels. With a small number of subjects in each group (four), significant differences in stress responses as measured by traditional physiological measures, or by voice stress analysis, or both, were not detected. All results, however, showed movement in the appropriate directions. (Diazepam reduced the degree of stress as indicated by both types of measurements.)

Smith (1973) analyzed fifteen hospital outpatients who suffered from general anxiety states. They were asked to respond to questions dealing with common life stressors and to three phrases about personal life stressors. The subjects were asked to reread the ten items for test reliability. It was correctly hypothesized that there would be a difference between the patient's self-report and the voice stress analysis. Smith concluded that the voice analysis method of identifying particular areas of anxiety would significantly enhance psychological evaluation and treatment.

Further work by Smith demonstrated that voice stress analysis could be used to reveal significant differences in anxiety states. One experiment contrasted voice stress data of professional broadcasters and of laypersons who called in to a radio talk show. Using both visual schemes and objective rating schemes to analyze voice stress traces, Smith investigated ten responses for each of twenty-two professionals and thirteen members of the public. It was found that the public's responses were significantly more stressed than those of broadcasters (Mann-Whitney U-test, p = .001).

In an experiment with eighteen neurotic outpatients (nine phobic, nine nonphobic), Smith assumed that the phobic group would respond to the experimental situation with a higher anxiety level than the nonphobics (obsessionals, hypochondriacs, and mild depressives). A control group of fifteen student nurses and professionals was also analyzed. Each subject counted aloud from one to ten and was instructed that the experimenter was concerned with anxiety levels in his voice. For the visually rated data, it was found that phobics were significantly more anxious than nonphobics (Whitney, p = .01). However, this experiment did not yield any significant differences between the nonphobics and the control group. Finally, to measure reliability for both the visual ratings and objective scoring methods, 368 subjects from both experiments were used in a split-half reliability test. By the Brown-Spearman formula, r = .39 for visual rating and r = .61 for objective scoring. The correlation between both scoring methods was highly significant (Pearson's r = .61).

The dental profession has long been concerned with the effects of stress during treatment. Johnson (1978) divided eighteen dental professionals into four groups based upon their members' amount of clinical experience and education in pedodontics. Slides of pedodontic situations of varying difficulty were shown to the subjects as stressors. Content in the slides included traumatic injuries and congential malformities. The nonstressing slides were simply textures of various colors that contained no cognitive content. The slides were projected onto a screen, and the subject was asked, "Do you see the slide?" The subject's "yes" response was recorded and then analyzed with voice stress techniques.

Johnson hypothesized that the least experienced groups would be the most stressed by the stressor slides and that practicing pedodontists would be the least stressed. He found the opposite. Analyzing his results, Johnson suggested that the higher stress exhibited by the experienced dental professionals came from their better understanding of the ramifications of what they were shown; thus, they experienced the highest levels of psychological stress. The conclusion of this study was that voice analysis supplies a valid indicator of stress. Johnson's work is consistent with previous research involving the physiological monitoring of anxiety states among experienced and inexperienced parachute jumpers (Tanner, 1976, pp. 26–27).

Worth and Lewis (1975) conducted a study in a dentist's office, utilizing the dentist himself as a stressor. While they did not find significant differences between the answers given on critical questions from a question list about brushing teeth in front of the dentist and answers given without the dentist, they did find significantly higher stress levels (chi square, $p. < .02$) on the answer to the first question ("Is your name _____?") when the dentist was present. This might again be explained by Reeves's (1976) finding that the effects of a stressor quickly fade.

Voice stress analysis seems to be most clearly validated in studies where the subject is somehow closely tied to the activity that produces the stress, when motivation and centrality are high. These conditions appear to relate most clearly to the phenomenon of elite political behavior. Indeed, it is somewhat surprising that the social sciences have not been more sensitive to the development of measures of remote assessment of communicative behavior. After all, much of what we describe as elite behavior is communicative behavior. Thus, a sharper focus on a more basic knowledge of communications could provide a deeper understanding of elite activity. Voice stress analysis may contribute to such an understanding.

Let us proceed, then, to the final voice stress analysis studies to be reviewed, politically oriented research. For the most part, these studies have been nonexperimental, and they have been performed in the context of international crises. It is important to point out that the crisis milieu was perhaps the safest area in which to begin politically oriented work, since we can make a reasonable assumption that leaders are indeed functioning under conditions of psychological stress. Furthermore, the issues with which they are dealing are of central concern to them because of the high positions of responsibility that they occupy.

The first published work to employ voice stress analysis to study crisis (Wiegele, 1978) examined the key television and radio speeches of Presidents Truman, Kennedy, and Johnson during the Korean, Cuban, Berlin, Tonkin, and Pueblo crises. A single, critical speech for each crisis was analyzed. What were some of the findings?

Each of the speeches expressed the sentiment that the U.S. was determined to see its way through to a successful resolution of the crisis. We called these determination themes. In crises in which there was a high probability of war (Korea, Berlin, and Cuba), the determination themes displayed very high stress. However, in crises in which war did not appear imminent (Tonkin, Pueblo), determination themes were substantially lower in stress.

Typically, crisis speeches contain references to a precipitating act or set of circumstances that led to the crisis situation. We subjected these precipitating act themes to voice stress analysis and found that such themes were in all likelihood viewed by presidents as givens of the crisis and that such themes were therefore low in stress value. In only one of our crises, the Gulf of Tonkin, was the precipitating act a theme of high stress. Our explanation of this was that the fact of a North Vietnamese attack in the Gulf was not established beyond doubt when Lyndon Johnson delivered his key address. Consequently, the precipitating act theme was apparently discomforting to the President, thereby resulting in high stress.

This exploratory study of only five cases had still another limitation. During international crises, presidents typically engage in a good deal of communication with their populations. This is the case because crisis demands not only concerted national action but also a reasonably continuous flow of information to a population to enhance their support for the leadership. Such support in turn contributes to the credibility of the positions advanced by the leader.

As a result of these shortcomings, we designed a study (Wiegele, 1980) that viewed presidential communications during a crisis on a longitudinal basis. Three crises were selected for analysis: Berlin (1961), the Dominican Republic, and the Cambodian incursion. The time frame for each crisis — fourteen, five, and six months, respectively — was expanded to allow for an extended voice stress analysis. Altogether, we examined dozens of audio recordings, which in printed form amounted to more than a thousand column-inches of text.

With regard to key crisis speeches, our findings corroborated the earlier crisis study, but further conclusions were also of interest. For example, we found that it was indeed possible to track a president's stress level over the extended course of the crisis and that that level bore some relationship to the objective state of affairs of the crisis itself. We also found that the more control that a leader perceives himself as having over events during an international crisis, the less stress he will exhibit in his voice. In some cases, we noted that careful voice analysis will be able to discern the leakage of important underlying clues or tipoffs to decisional behavior that could strongly influence subsequent policy choices. An example of the latter involved President Johnson's references during the Dominican crisis to U.S. civilians who were "trapped" on the island while events proceeded. These references produced the highest stress values of the crisis for Mr. Johnson. While an analyst might have hypothesized that more substantive issues directly related to the crisis itself would have caused the highest stress in the President, that was not the case. Our interpretation of this apparent anomaly was that Mr. Johnson was an immensely political man who viewed the domestic impact of unprotected U.S. civilians with grave discomfort, thereby increasing his stress levels on this topic.

Brief mention should be made of one other political voice-stress analysis study. Although it did not deal with international crisis, this research effort (Wiegele, 1978) examined the vocalizations of Richard M. Nixon at two stress-laden moments of his public life: his concession speech upon losing the California gubernatorial race in 1962 and his farewell to his staff after he resigned from the presidency in 1974. At the conclusion of that study we argued that "events that are traditionally viewed by the electorate as political failures . . . were perceived by Nixon as serious personal failures. Indeed, in the two speeches that were analyzed, Nixon never addressed the broader political consequences of the occurrences" (Wiegele, 1978, p. 74). When Nixon alluded to politics in these speeches, he displayed little stress; when he talked about personal failure, his stress levels were exceedingly high.

Methodological Concerns

There are three significant areas of methodological concern for voice stress analysis: conscious control of vocalizations, alternative factors causing stress, and coding. The latter will be addressed in detail after a brief discussion of the first two validity questions. (For a fuller discussion, see Wiegele, 1980).

Brenner (1974) found that subjects given knowledge of voice stress analysis were able to conceal deceptions from an analyst. As noted above, this may be due to the fact that the instrumentation can measure stress only at the time of vocalization. Thus, it may be possible to use certain split-second techniques to raise or lower arousal levels more or less at will; for example, by taking a deep breath, coughing, or clenching one's teeth or hands. Although this finding could seriously limit the use of voice stress analysis as a lie detection device, it should not have much effect on more remote uses on naive subjects in analysis of psychiatric exams, group meetings, or nondeception experiments or speeches.

Another virtually unexamined source of concern is the alternative factors question. The only published mention of this is by Brenner, Branscomb, and Schwartz (1979), who found that, on repeated vocalizations of the numbers zero to nine, the stress scores on the numbers were not random as expected. Rather, there was a hierarchy of stress scores, ranging from five (highest) through nine, four, one, zero, seven, three, and two to eight (lowest). (Six, as vocalized, did not afford a large enough trace for analysis.) Is it possible that vocalizing certain vowel sounds (like ī) affects the physiological basis of the voice stress analysis traces? If so, is it substantial enough to skew an analysis?

Along these lines, we have found that the loudness of speaking affects the traces, especially when a person is shouting. We have termed these trace configurations *tight* and we hope to do further analysis to determine whether tightness is habitually coded as stressful. Other concerns that might affect voice patterns are speed of speech, accent, hoarseness, and illness. However, the possible effects of such factors is conjectural at this point.

Finally, there is the problem of coding. All critics of voice stress analysis point to the subjectivity of the coding and the resulting low intercoder reliabilities as indicators of the "softness" of voice analysis. Analysts have used anywhere from a five-point scale to a two-point stress–no-stress scale (Wiegele, 1978). Reported intercoder reliabilities have ranged between .38 (Horvath) and .56 (Worth and Lewis) when reported. Usually, a score of .80 is considered necessary for reliability.

Since such low intercoder reliability scores are unacceptable, we attempted to devise an objective coding scheme that would be reliable and valid. Building from the assumptions of the physiology of the PSE as related below, we developed manual coding procedures that can give a numerical (interval scale) score for any word. We view this method as preferable to the method of Smith (in press), which seems not only tedious but also open to significant error in certain circumstances (for example, when only one or two pulses in a trace register over the midpoint of the total span of deflection).

The procedure that we developed relied on the Dektor Company's concept of center of mass (of the recorded pulse). For each pulse in a trace, a middle point could be marked, and these points could be connected by a line that would indicate both the degree of slope(s) and the number of slope changes. Unfortunately, while some aggregation of slopes within a trace is possible, the mechanical difficulties of this type of analysis are enormous. One must measure the angles for all slope changes and average them. Questions of whether to start at the bottom (thus producing large first angles), the third point, the second quarter, and so forth, have to be answered, and all have to be answered subjectively. To avoid this tedium and subjectivity, we decided to use the top points of the pulses (each a discrete excursion of the recording pen) for our analysis.

The rationale for voice stress analysis is that the fluctuations of the pulses in a trace should dampen as stress increases. Using this logic, we hypothesized that the number of slope direction changes (with control for the length of the trace) would be an indication of the degree of frequency modulation. A blocked pattern (straight across, zero slope) indicates the highest stress, and so the no-slope condition between pulses must also be analyzed.

A simple counting procedure was devised to account for both conditions. Each positive or negative inflection in slope is scored as 1; each "no change" is scored as −1. The arithmetic sum of the scores is then expressed as a proportion of the total number of pulses scored. For example, if there were 19 pulses for a given word, 6 of which were inflected and 2 of which were not, the score would be $6 - 2 = 4$. If the total number of pulses were 19, the score would be 4/19 or .21, which indicates mild stress. Figure 3 illustrates this example.

We then developed explicit rules for coding and tested for the reliability of this coding scheme. Five voice stress analysis coders trained in subjective coding were asked individually to rank one member each of 300 pairs of traces (twenty-five words in all combinations) as the more stressful. Thus, for each coder we could rank the twenty-five words on a continuum, test for within-coder reliabilities, and compare intercoder ranking scales.

Within-coder reliabilities were derived by Guttman scaling of the paired rank decisions. A perfectly consistent coder would receive a coefficient of reproductibility of 1.00. Our coders ranged between .82 and .93. While this is high, it shows enough within-coder inconsistency to create concern for the analyst. Intercoder reliabilities were derived from Pearson correlations of Guttman scores for each coder. These correlations ranged between .69 and .88, substantially better than those reported for other scoring methods.

"Objective" scores were calculated for each word according to our procedure and correlated with the coders' Guttman scores. These correlations ranged from .59 to .74 (all significant at .002); these are lower than the intercoder reliability socres. Upon examination and discussion of the traces and the differences in the scoring, however, we were satisfied that the objective coding scheme was valid and that personal biases in the subjective coding accounted for a large percentage of the unexplained variance. The intercoder reliability score for our objective coding system was .948, and we felt that this afforded a highly consistent basis for analysis.

Figure 3. Slope Direction Change in One Trace (Expanded)

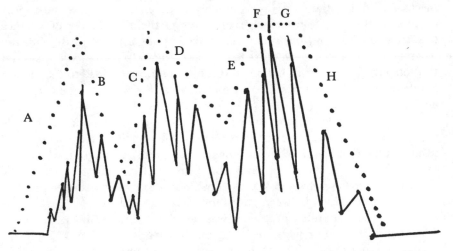

There remain various problems in the reliable scoring of traces (for example, short traces, monosyllabic versus polysyllabic words, tight design, words that run together, cluttered bottom trace due to stylus restriction), but we are encouraged by the method's ability to discern the stress level objectively. While the theoretical range for the scores is − 1 (total stress) to 1 (complete nonstress), the working range that we have found is − .15 to .60, with the mean around .30. Because our data have been derived from presidential speeches on crisis matters, the mean is probably lower (that is, more stressful) than in normal speech patterns.

Conclusions

At this point in its development, voice stress analysis must be considered to be an emerging technology of potentially great research applicability. As Lykken (1974) observed, voice stress analysis is an important addition to the repertoire of physiologically based research techniques. The useful applications to which voice stress analysis may be put in sociopolitical research are only beginning to emerge. In the body of PSE literature that has already been produced, it is clear that voice stress analyses have provided impressive results in real situations involving real challenges to an individual's role, concerns, or situation. In experimental situations where controls over the intensity of the stressing stimuli have been nonexistent or where no real challenges exist, results with voice stress analysis have been mixed.

It appears that if social analysts want to employ the voice stress methodology in their research concerns, it is they who will have to perform both the experimental and the applied validity studies that are so urgently needed. Given the state of professional training of most social and political scientists and their reluctance to adopt physiologically based methodologies, it seems unlikely that such work will be done by them alone. Let us hope that interdisciplinary teams of researchers will confront this problem in the near future. If answers to the methodological questions that we have raised are forthcoming, voice stress analysis has the potential to become a standard tool of political research.

References

Bell, A. D., Jr., Ford, W. H., and McQuiston, C. R. "Physiological Response Analysis Method and Apparatus." Canadian patent no. 943230, issued March 5, 1974, and U.S. patent no. 3, 971, 034, issued July 20, 1976.

Borgen, L., and Goodman, L. "Audio Stress Analysis, Anxiety and Anti-Anxiety Drugs: A Pilot Study." Unpublished manuscript, no date.

Brenner, M. "Stagefright and Stevens' Law." Paper presented at meeting of Eastern Psychological Association, April, 1974.

Brenner, M. Branscomb, H., and Schwartz, G. E. "Psychological Stress Evaluator — Two Tests of a Vocal Measure." *Psychophysiology,* 1979, *16* (4), 351–357.

Horvath, F. "An Experimental Comparison of the Psychological Stress Evaluator and the Galvanic Skin Response in Detection of Deception." *Journal of Applied Psychology,* 1978, *63* (3), 338–344.

Johnson, J. B. "Stress Reactions of Various Judging Groups to the Child Dental Patient." Unpublished master's thesis, University of Iowa, 1978.

Lewis, B. J., and Worth, J. W. "Transfer of Stress Through Verbal and Written Communication." Unpublished manuscript, 1975.

Lykken, D. "Psychology and the Lie Detector Industry." *American Psychologist,* 1974, *29* (10), 725–739.

McGlone, R. "Test of the Psychological Stress Evaluator (PSE) as a Lie and Stress Detector." Paper presented at Carnahan Conference on Crime Countermeasures, Lexington, Kentucky, 1975.

McGlone, R., and Hollien, H. "Partial Analysis of Acoustic Signal of Stressed and Unstressed Speech." Unpublished manuscript, 1976.

Reeves, T. E. "The Measurement and Treatment of Stress Through Electronic Analysis of Subaudible Voice Stress Patterns and Rational-Emotive Therapy." Unpublished doctoral dissertation, Walden University, 1976.

Shearer, W., and Wiegele, T. C. "A Comparison of Vocal Stress Analysis and Skin Responses." Paper presented at annual convention of the American Speech and Hearing Association, Chicago, November, 1977.

Smith, G. "Voice Analysis for the Measurement of Anxiety." *British Journal of Medical Psychology,* in press.

Smith, G. "Analysis of the Voice." Unpublished manuscript, 1973.

Smith, G. "Lie Detection by Voice Analysis: The Problem of Reliability." Unpublished manuscript, no date.

Tanner, O. *Stress.* New York: Time–Life Books, 1976.

Wiegele, T. C. "Decision-Making in an International Crisis: Some Biological Factors." *International Studies Quarterly,* 1973, *17* (3), 295–336.

Wiegele, T. C. "Health and Stress During International Crisis: Neglected Input Variables in the Foreign Policy Decision-Making Process." *Journal of Political Science,* 1976, *3* (2), 139–144.

Wiegele, T. C. "Models of Stress and Disturbances in Elite Decision-Making." In R. S. Robins (Ed.), *Psychopathology and Political Leadership.* New Orleans: Tulane Studies in Political Science, 1977.

Wiegele, T. C. "The Psychophysiology of Elite Stress in Five International Crises: A Preliminary Test of a Voice Measurement Technique." *International Studies Quarterly,* 1978, *22* (4), 467–512.

Wiegele, T. C. "Remote Psychophysiological Assessment of Elites During International Crises." Report prepared for the Cybernetics Technology Office, Advanced Research Projects Agency, January, 1980.

Wiegele, T. C., and Plowman, S. "Stress Tolerance and International Crisis: The Significance of Biologically Oriented Experimental Research to the Behavior of Political Decision-Makers." *Experimental Study of Politics,* 1974, *8* (3), 63–92.

Wiggins, S. L., McCrainie, M., and Bailey, P. "Assessment of Voice Stress in Children." *Journal of Nervous and Mental Disease,* 1975, *160* (6), 402–408.

Worth, J., and Lewis, B. "Presence of the Dentist: A Stress-Evoking Cue?" *Virginia Dental Journal,* 1975, *52* (5), 23–27.

Leonard Hirsch is a doctoral student in international relations and international political economy at Northwestern University.

Thomas C. Wiegele, director of Northern Illinois University's
Center for Biopolitical Research, has published widely on
voice stress analysis. His monographs include Biopolitics:
Search for a More Human Political Science. *(1979).*

Index

Addendum to *New Directions for Methodology of Social and Behavioral Science,* No. 6, 1980